Praise for Thomas I
The Christian Futur

D0862934

"Mary Evelyn Tucker an   .cious
gift in bringing together these ten key essays by the late
ecologian, Thomas Berry. These essays distill the basic
elements of his critique of Western thought and culture and
the transformation needed to bring humans and the earth
into harmony for mutual survival and flourishing."

—Rosemary Radford Ruether
Claremont School of Theology
and Graduate Theological Union

"Today, Christians are torn between two visions of themselves.
One is that of segregated people, cut off from the animals and
plants, people of other religions, and overly-preoccupied with
questions of personal salvation. The other is that of people
who humbly recognize that they—we—are small but included
in a larger and evolving whole, and that our lives are engraced
by other creatures, by people of other religions, and by the
stars and planets. In the twenty-first century no thinker has
called us toward this second vision more consistently, more
profoundly, and more beautifully than Thomas Berry... There
is good news to be heard. Its promise lies in the pages of this
book and the vision of its author."

—Jay McDaniel
Willis T. Holmes Distinguished Professor of Religion
Hendrix College

"These essays by Thomas Berry are special in their focus on
the deep connections between Christian theology and the
universe story. One of the deepest is that *both* underscore the
need for sacrifice in our time of climate change and economic
meltdown. Berry reminds us that 'the wisdom of the cross and
the wisdom of the universe' mean 'that sacrifice with each

other and for each other is a dimension of life itself' (94). This is a message we must hear and obey!"

—Sallie McFague
Distinguished Theologian in Residence
Vancouver School of Theology

"Much gratitude to Thomas Berry, who calls us to awaken to the living universe as the primary religious, physical, material reality. He stresses that humanity is a mode of being of the universe as well as a distinctive subject in it. Only when we realize this can religion, especially Christianity, contribute to the religious and ecological demands of our era. Berry critiques the Christian preoccupation with the limits of the 'human condition' and its narrow religious horizon. Yet he offers ways to resituate and reinvigorate key Christian themes—the wisdom of the cross, the universe as cosmic liturgy—to enable Christians to fulfill the tasks of the twenty-first century. There is no future for Christianity outside of the well-being of the Earth!"

—Heather Eaton
St. Paul University

THE CHRISTIAN FUTURE
AND THE FATE OF EARTH

ECOLOGY AND JUSTICE SERIES

THE CHRISTIAN FUTURE AND THE FATE OF EARTH

By Thomas Berry

Edited by Mary Evelyn Tucker and John Grim

ORBIS BOOKS

Maryknoll, New York 10545

Library of Congress Cataloging-in-Publication Data

Berry, Thomas Mary, 1914–2009
  The Christian future and the fate of earth / by Thomas Berry ; edited by Mary Evelyn Tucker and John Grim.
    p. cm. — (Ecology and justice series)
  Includes index.
  ISBN 978-1-57075-917-8 (pbk)
  1. Human ecology—Religious aspects—Christianity. 2. Ecotheology. I. Tucker, Mary Evelyn. II. Grim, John. III. Title.
  BT695.5.B468 2009
  261.8'8—dc22
                    2009008811

To all the wonderful women who,
assisted by my writings,
have committed themselves
to the Great Work of Earth Community
and the revelatory nature of the universe

Contents

Preface

John B. Cobb, Jr.

By Earth Day 1970 it had become clear to a good many of us that one danger loomed over all others for the whole of humanity and indeed the whole earthly biosphere. For the first time in human history, humanity as a whole—and not simply local groups—was threatened by destruction. We had earlier assimilated the fact that a nuclear war could have such consequences, but now we saw that even if such a catastrophe were avoided, the danger remained. It was not that someone might take a peculiarly dangerous or extraordinary action. It was rather that the consequence of continuing in the ruts in which we were already travelling would carry us over a precipice.

When this realization struck, most of us were already involved in multiple causes. We were trying to support the Civil Rights movement and stop the war in Vietnam. There was a liberation movement in Latin America, and feminism was gaining attention in the United States. Poverty remained a terrible problem worldwide. There were injustices everywhere. The response of many people was to add the "environmental crisis" to this list of important concerns. But it was obvious to others of us that this was not enough.

My own initial response was to emphasize the priority of making the changes that could lead to survival. That involved a certain withdrawal of time and energy from other causes as well as making statements that seemed to belittle other concerns. But after a few months, I changed my mind and followed the leadership of the churches. Those in the church who understood the situation decided that what was needed was to build alliances between environmental, peace, and justice-oriented groups. The World Council of Churches added "sustainable" to its characterization of the good society as "just" and "participatory." In the United States many church-related groups called for "eco-justice."

This strategy reduced the suspicion on the part of ethnic minority groups and Third World peoples that improving the environment was just another excuse being employed by the privileged for avoiding basic issues of justice. In addition, it encouraged the efforts of different people to find particular handles on environmental issues around which wide consensus could be reached. But this strategy also allowed the great majority of concerned people to treat the issue of sustainability as simply one more to be put on the laundry list. In the local church the issue often received attention just one Sunday in the year. Few could claim that as a result of the widespread recognition of the importance of developing a sustainable society the church had reordered its priorities so as to deal effectively with what is clearly the most important of all the problems we face.

One Christian did not allow himself to be diverted by other issues. His name is Thomas Berry. He knew that humanity as a whole faced its ultimate crisis. Every issue should be viewed in light of this fact. Its claim to priority should be fully acknowledged. Even those of us who have tried to weave the concerns of peace, justice, and sustainability together are profoundly grateful that his witness to priorities was not diluted.

Others of us saw how important it was to work at many levels. We needed improved technology and more efficient use of resources. We needed businesses to change their practices. We also needed different economic theories and practices. We could not get these without political changes, so we needed these as well. We recognized that a basic change in attitude and vision on the part of all of us had to take place, but this was just one level among the many that called for our attention.

Berry believed that the changes we need will not occur at the many levels until they occur at the basic one—the way we understand ourselves and our world. Again he refused to be distracted from the fundamental task. Further, his work in the history of religion led him to believe that the key to changing the way people see themselves and their world can be found in their creation stories. He judged that today such a story must correspond with what is known scientifically. But simply repeating what physicists and astronomers say does not meet the need. The universe must be understood as a communion of subjects, and the story of its rise and development must be told in these terms. Such an understanding would lead to a sense of participation in the universe and in the present stage of the story. With the help of Brian Swimme he wrote a truly remarkable account of the ongoing creation of which we are a part, *The Universe Story*. Building on that account he named the next phase of life on this planet, "the Ecozoic Age." By using the language of geological epochs, he made clear the radical uniqueness of this crisis. He also communicated hope that the required changes would occur.

Thousands of people, perhaps tens or even hundreds of thousands, have been led to give real primacy to the task of living into the Ecozoic Age. No other writer in the ecological movement has had analogous effectiveness. In the decades ahead, more and more people, tens of millions at least, will fully recognize that the ecological crisis has the ultimacy that Berry has insisted on throughout his career. Others will come up with new formulations and make different proposals. But Berry's formulation has pride of place, and it may prove the most durable and effective of all. However that may be, we all have reason to be deeply grateful for his unique calling and his extraordinary faithfulness to it.

Introduction

Mary Evelyn Tucker and John Grim

This collection of essays written over the last several decades represents Thomas Berry's vintage reflections on the contributions Christians can make to a common future for the planet. He outlines both the problems facing Christianity in this regard as well as the promise. In doing so, he brings his reader into a place of possibility—a new grounding for both reflection and action. Written by one of the foremost thinkers of our period, these essays are not only timely, they are vital for finding our way forward in this new millennium.

With Thomas Berry's death on June 1, 2009, these essays take on a new sense of importance. While many people have followed his writings on the story of evolution, few have understood his Christian roots. He remained a priest, teacher, and writer throughout his life. Berry expanded the framework of his Catholic heritage by first studying other religions, especially Asian and indigenous traditions. Later, drawing on the French Jesuit and paleontologist Pierre Teilhard de Chardin, Berry enlarged his scope of concern to include the universe and Earth. Like Teilhard, he situated the human as arising from the evolutionary process. For Berry, the human is the self-reflective dimension of the universe. Thus, with his study of human history and of Earth history, Berry opens up Catholicism to be present to all of creation and to be concerned for its future.

Thomas Berry was a person of immense feeling for the suffering dimensions of life. Early on he would visit Dorothy Day and Peter Maurin at the Catholic Worker in the Bowery. He was a firm supporter of social justice issues. However, as this Orbis Books series in ecology and justice demonstrates, he wished to re-envision not only equitable human-human relations but also life enhancing human-Earth relations. Without this, a sustainable future is not possible. Leonardo Boff and others have taken Thomas Berry's vision to heart. A vibrant Earth community requires a new awakening to the incarnational dimension of life—human and natural.

Berry does not minimize the size and scale of the challenges we are facing as a species and as a planet. The urgency of the global environmental crisis is now well documented, even if its many manifestations are not yet fully understood. From climate change to pollution of air, soil, and water, from population growth to biodiversity loss, Berry is aware of the varied problems included in the broad term "environmental crisis." What is particularly remarkable is how early Berry understood the magnitude and complexity of these issues. While most people ignored his warnings over thirty years ago, no one can afford to now. He is no longer a lone prophet calling out in the wilderness. Thus his writings take on a special relevance at this stage of our search for new and sustaining human-Earth relations. Indeed, the noted theologian John Cobb observes in his preface, "No other writer in the ecological movement has had analogous effectiveness" in helping us realize the "radical uniqueness of this crisis."

The Challenge and the Promise

Consistent with this spirit, Berry strongly challenges Christianity to respond appropriately to this complex situation. He raises penetrating questions regarding some of the reasons, both historical and contemporary, for the lateness and laxity of the Christian response. Is this because the desire for personal salvation into a heavenly realm supersedes all other concerns? In other words, does the search for otherworldly rewards override commitment to this world? Has the concern for Redemption lessened the concern for Creation? Has the material order of nature been devalued by Christianity? Have human-centered ethics been all consuming so that we do not have an ethics that addresses ecocide and biocide? Did Christianity surrender its interest in natural theology and cosmology to positivistic science? Are seminaries—both Protestant and Catholic—so preoccupied with an outdated curriculum and sectarian theological issues that they have no room to include environmental concerns? These questions require even further reflection, he suggests, before an adequate response to our situation can be formulated from out of the Christian tradition.

Yet, he observes, the promise inherent in Christianity is still to be fully acknowledged and expressed. This includes the recognition in Paul's epistles regarding the Incarnation as embracing both the historical person of Christ and the Cosmic Christ of the universe. Berry also speaks of the deep appreciation for the order and beauty of Creation contained in the Christian tradition ranging from Francis of Assisi and Thomas Aquinas in the medieval period to Pierre Teilhard de Chardin in the twentieth century. He writes often of the liturgical cycle as grounding humans in the great seasonal changes of nature. He notes as well the powerful material reality of the elements of nature used in the sacraments such as bread and wine in the Eucharist and water in baptism. He has wondered what it means to baptize with water if we have polluted it, or to receive bread and wine fashioned from wheat and grapes laced with pesticides and herbicides.

In addition, he understands the immense promise in the emerging alliance of social justice and environmental justice that is being forged, especially in a book series such as this one produced by Orbis. As we would gather over the years to discuss the news of the day, he would often comment on the suffering the environmental crisis is inflicting on those most vulnerable. Ever since the toxicity of Love Canal near Buffalo, New York, was exposed in 1978, environmental health issues were evident in his concerns. It also became clear to him that many toxic waste sites in the United States and abroad are frequently placed near poor communities. In more recent years he would comment on the deleterious impact of climate change on island nations and coastal peoples such as the millions in Bangladesh and in New Orleans. He wondered frequently how new human-Earth relations could be forged that would be mutually beneficial, not harmful for the whole Earth community.

For Berry some of the key sources of transformation—theological, ethical, ritual—were within the religious traditions waiting further articulation and action. He was hopeful that Christians might respond more confidently and cogently to growing environmental threats to people and the planet. Indeed, he opened up a gateway for Christians to reform their tradition by making an Exodus passage into the modern world. He called for a transition like the first Exodus experience of the Jews out of Egypt, a passage into modernity, something the Church has so often resisted. In critiquing the ineffectual response of Christian institutions—both academic and clerical—to the environmental crisis, he also suggested that it is not too late. Indeed, it is more important than ever before that these institutions become involved. He invited theologians and laity alike to make their contributions—in universities, in seminaries, in churches, and in grassroots movements.

Christians Responding to the Environmental Crisis

It is significant to note that Christianity is already shaking off its lethargy with regard to environmental issues and many denominations are making efficacious responses. For more than a dozen years, the Protestant-based World Council of Churches has published treatises on "justice, peace, and the integrity of creation" and emphasized the ethical dimensions of climate change. For fifteen years the Ecumenical Patriarch Bartholomew, leader of the Greek Orthodox Church, has organized major religion and science symposia on water issues across Europe, as well as in the Amazon River basin and in Greenland. He has issued statements—including a joint statement with John Paul II in 2002—calling destruction of the environment "ecological sin." The Evangelical community in the United States has published position papers calling for care for creation and warning that the poor will suffer from global warming. The U.S. Catholic bishops have issued a letter on global climate change as well as other statements on ecology, including a letter on the Columbia River bioregion. Pope Benedict XVI has warned frequently that the beauty and integrity of Creation is being severely diminished by humans. Moreover, his encyclical *Charity in Truth* (*Caritas in Veritate*) is a strong statement condemning the expoitation of land and people under the current economic system.

Drawing on Berry's ideas, the Catholic bishops of the Philippines issued in 1988 a pastoral letter on the environment titled "What Is Happening to Our Beautiful Land?" Two decades later, in December 2008, they published another statement listing the critical environmental problems their country is still facing and calling for a moratorium on mining and logging. In February 2009 the Catholic bishop of Alberta, Canada, wrote a strong condemnation of oil extraction from the tar sands, noting that such widespread environmental destruction is morally reprehensible. All of these statements are being used as ethical calls to engage Christian communities in further action on behalf of the environment. These statements, along with those from other religious traditions, have been collected on the Forum on Religion and Ecology website (www.yale.edu/religionandecology). A website has been created for teaching Christianity and ecology in seminaries, colleges, and churches (www.webofcreation.org). In addition, a joint degree program between Yale Divinity School and Yale School of Forestry and Environmental Studies has been developed.

However, our desire for both reflection and action, Berry observes, may require even deeper contemplation of the roots of the problems. This is why Berry brings us toward the universe story as a comprehensive context for responding to our ecological role in the modern world that is being ravaged by industrial processes. He feels that this context of evolution will evoke awe, wonder, and humility in humans. At the same time, as a functional cosmology, it will encourage the "great work" of ecological restoration, ecological economics, and environmental education so needed in our times.

He notes how the Christian tradition in many of its key doctrines, such as the Trinity, Incarnation, and Redemption, can be understood within this large cosmological framework. These essays where he makes linkages between Christian theology and the universe story have hitherto not been available for a wide audience. His comments are richly suggestive and provide fertile ground for theologians and laity alike. They illustrate Thomas's broad appreciation of the Catholic tradition and his indebtedness to Thomas Aquinas after whom he took his religious name.

Thomas Berry as Teacher, Scholar, and Mentor

Thomas's reflections on the *Christian Future and the Fate of Earth* arise from a long-term commitment to the Catholic tradition as a monk, scholar, teacher, and writer. He became a member of the Passionist Order during his college years and many of his ideas about sacraments and liturgy were formulated out of the rhythms of monastic life. This contemplative life allowed him time for reading and thinking, time which he avidly sought, often reading into the depths of night. It was as if he was driven to see and understand a larger vision of human life in this most critical of moments for the planet.

After completing his doctorate from Catholic University with a thesis on Giambattista Vico, he went to study in China in 1948–1949. There he met Theodore de Bary who was to become a lifelong friend and one of the most renowned Asian scholars in the West. Along with his wife, Fanny de Bary, they were Thomas's earliest supporters. Over frequent dinners at their home in Tappan, New York, Thomas and Ted would discuss the spiritual dimensions of the Asian classics, especially Confucianism. Fanny shared Thomas's interest in Teilhard and for many years was an active board member of the American

Teilhard Association, including the decade during which Thomas served as president. Drawing on her years of study of Japanese flower arrangement, she always brought beautiful spring flowers to the annual Teilhard Association meetings in New York. Her feeling for beauty in nature was matched by Thomas's aesthetic sensibilities. At Columbia University Ted established a groundbreaking Asian studies program highlighting the classical texts, history, and culture of India, China, and Japan. He assisted in having key Asian classics translated and published by Columbia University Press in monographs and in sourcebooks. Thomas used these books in his teaching and writing, as did many professors and students in Asian studies across North America. Ted also founded the Oriental Thought and Religion Seminar at Columbia, which he invited Thomas to join. It was a rich and sustaining friendship with Ted and Fanny when few understood Thomas's keen interest in Asian religions or appreciated his fascination with the evolutionary story that Teilhard had articulated.

Thomas began his teaching at Seton Hall University in New Jersey and then moved to St. John's University on Long Island. The Jesuit Christopher Mooney later invited him to join the Theology Department at Fordham University. There Thomas founded and directed the History of Religions program for more than a dozen years before retiring from teaching in 1979. This was the only program of its kind at any Catholic University in North America. Regrettably it did not endure beyond his tenure at Fordham. Yet many of his students continue to teach at universities across the United States and Canada. The Passionist theologian Steve Dunn, who had studied with Thomas, directed a doctoral program at the University of Toronto. There he trained a generation of scholars in Berry's perspective on the universe story and ecology.

Thomas was somewhat anomalous in Fordham's Department of Theology. He was not a Jesuit, nor was he a theologian. Instead, he was trained in Western history and in the world's religions. Yet, he was a charismatic figure and in demand as an engaging speaker. Thus, the history of religions section that he directed eventually drew more students than any other program in the Theology Department. Students came from around the country, some turning down admission to religious studies programs at Columbia or Yale, to study with him. And what lively, dedicated students they were!

This is where we met in the fall of 1975—John coming from North Dakota and exploring Native American religions, Mary Evelyn recently returning from Japan and immersed in the Asian religious traditions. It was an exhilarating time for us as we gathered with other graduate students to study with this original thinker and incomparable mentor. We thrived on the challenges he presented to us: learn the textual language of at least one tradition, know the history of many, feel the bass notes of the spiritual wisdom of each tradition, and explore history and culture so that the living context of a tradition might open up.

Thomas himself set a high bar for his students, having read widely in the world's religions and learning the languages needed to appreciate their ancient texts and commentaries. His library at the Riverdale Center for Religious Research reflected this passion for breadth and depth. The Center was a beautiful old Victorian house overlooking the Hudson River just outside New York City. We gathered there under the spreading limbs of a four-hundred-year-old red oak and across from the 200-million-year-old rock cliffs of the Palisades. Over twenty-five years friendships were formed, talks were held, and lasting memories were made at this incomparable center. During the academic year we would have monthly lectures on Saturday followed by potluck dinners. In the summers Thomas would hold a conference on a topic of his choosing. For weeks in advance the conference room at the Center was covered with his books and notes as he prepared his lectures. Then several hundred people would gather for a weekend at the retreat center next door to hear his latest thinking. These were inspiriting moments!

Thomas had collected his library of some eight thousand books in this rambling and accommodating house. In the large front room the Latin Church Fathers faced the Greek philosophers and the Chinese classics, all in their original languages. The Sanskrit texts of the Hindu classics were in the next room and he initiated many of us into the *Bhagavad Gita* in Sanskrit before we read the other Indian texts. Some of his students wrote dissertations that were translations of these texts with original interpretive commentaries.

Not only did his library hold the classical scriptures of the Asian and Abrahamic religions, but also ethnographies of indigenous peoples, as well as collections on ecology and environmental issues. Upstairs overlooking the Hudson River was the American Indian room, filled from floor to ceiling with books on the various tribes that populated the North American continent. This was John's favorite haunt, as the sun porch was Mary Evelyn's. There on the porch we would join Thomas for lunch or dinner. Music such as Beethoven's Archduke Trio was background for wide-ranging conversations from the state of the world to the state of Sung China, Heian Japan, or Mughal India.

His historical versatility was stunning—unmatched by any professor we had ever encountered, with the exception of his colleague Ted de Bary. We were the beneficiaries of these two brilliant minds, as Mary Evelyn went on to Columbia to pursue her PhD in Confucianism with Professor de Bary. This was during the "Golden Year" of 1977 to 1978 when John lived with Thomas, Brian Brown, and Valerio Ortolani at the Riverdale Center. Thomas introduced and married several of his students to one another. Somehow they fell in love studying with him! So it was, that with Thomas presiding, John married Mary Evelyn in the midsummer of that golden year.

With this wedding, our four decades period of work with Thomas was launched. We were just beginning a journey with him that would stretch beyond our graduate years and into our teaching years. He was with us throughout, working with us and attending and speaking at many of the conferences on religion and ecology that we convened at Harvard in the 1990s. At the culminating conference in 1998 at the American Museum of Natural History in New York, more than one thousand people had gathered. They were so enamored with his concluding speech that they would not let him leave the stage when the moderator indicated his time was up.

Thomas's understanding of the wisdom and the history of the world's religions was remarkable. Well before interreligious dialogue became a topic of inquiry and discussion, he was immersing himself in the texts and traditions of India, China, and Japan. He wrote a book titled *Buddhism* and another titled *Religions of India*; both books, published by Columbia University Press, are still in print. There was no question in his mind regarding the depths of spiritual and practical insight conveyed by these traditions. He would remark, in light of the Vatican II document *Nostra Aetate* that spoke of "rays of truth" in the world's religions, that, indeed, they held not just rays but floods of illumination and truth.

Thomas moved through diverse religious traditions with an empathetic feel for the pulse of their spiritual dynamics. We recall afternoons after class when a group of us would gather with him in the campus dining area or ratskeller. We would explore the Pali texts of Buddhism, the Sanskrit scriptures of Indian *dharma*, and the enigmatic turn of the hexagrams in the Chinese *Book of Changes* (*I Ching*). More than once he guided us through the process of forming a hexagram from this classical text in response to a question. Rather than dwell on the divinatory dimension of the *I Ching*, however, he urged us on to deeper reflection on the poetic lines of the text. We still ponder the possibility in our driven, acquisitive modern world that, as one hexagram indicated, "the small may get by" (*hsiao kuo*). Perhaps, he would observe, we may even move toward a world where Fritz Schumacher's notion, "Small is Beautiful" may come to be realized.

While those graduate school days focused on historical and textual developments in the many religions, Thomas encouraged us also to explore the cosmology of religions. Under his guidance we related rituals, texts, teachings, and commentarial studies to the stories of creation and metaphysical speculation about the world as it was and where we humans fit in. We struggled to discover the history, anthropology, and sociology embedded in those stories. Thomas forged ahead, articulating his broad understanding of historical interactions and cultural relationships.

Gradually, we came to see his pursuit of cosmology as the ground for reflecting on human meaning and orientation in the world. "With a story," he would say, "people can endure catastrophe. And with a story they can gather the energies to change their lot." For him the first place to look for story was in history. He began with Western history and later moved to Asian religions and Native American traditions. He was part of the early group of world historians seeking to define the contours of our human movement and exchanges across the planet. He mused that the West was in search of a comprehensive story and cited historians such as Oswald Spengler, Arnold Toynbee, Christopher Dawson, and Eric Vogelin to give nuance to his views. He drew on the philosopher of history Giambattista Vico for describing the sweeping ages of history—both human and Earth. In Vico he found a mind that could roam amidst the large-scale dimensions of earlier eras and through whom Thomas could uncover a passageway into modernity. It was because of his remarkable grasp of world history that Thomas could eventually make the transition into evolutionary history.

In his classes he would grope for a thought, searching for a word that could capture the transition between the great ages of evolutionary time. And, then, he would cough. That cough became emblematic for us of his search for articulation—looking for the words to move us forward into a new and deeper understanding of our historical moment. Gradually, Thomas connected his study of history and evolutionary cosmology to the environmental issues of our day. This came slowly, maturing like some fine wine that carries the texture and taste of soils, sun, grapes, air, and aging.

Reaching into his own past he recalled his boyhood experience in North Carolina of a summer meadow filled with white lilies. This experience began to define his commitment to preserve and protect such beauty. Increasingly he spoke of a deep affectivity and authenticity imparted by Earth itself in its biodiversity. It was in the early 1980s that these ideas coalesced in the notion that we were ending the Cenozoic era and entering an "Ecozoic" period. This was his way of naming the terminal destruction of Cenozoic life in the industrial-technological bubble of consumer acquisitiveness. He observed that scientists were telling us that we were in the midst of an extinction period. Nothing this devastating had occurred since the dinosaurs went extinct 65 million years ago. But rather than leaving his audience in despair, his use of the term Ecozoic was to name that emerging period in which humans would recover their creative orientation in the world.

Sensing his way forward, he drew increasingly on the thought of Pierre Teilhard de Chardin for insight into the story of our times, namely, the emerging, evolutionary universe. Teilhard provided a large-scale vision of humans as situated within the vast context of universe evolution. Teilhard had a profound sense of the unfolding of ever-greater complexity and consciousness in the arc of evolution from the molecular to the cellular to multi-cellular organisms to the explosion of life forms.

Rather than settling on Teilhard's insights, however, Thomas pushed beyond to explore the conjunction of cosmology and ecology. While appreciating Teilhard he also critiqued his optimistic view of "Building the Earth" with new technologies and scientific discoveries. He balanced Teilhard's faith in technology with a strong dose of realism—highlighting our current patterns of environmental degradation. He wanted us to see that in a geological instant we were extinguishing life—species, ecosystems, rivers, wetlands. Our contemporary moment was as significant as the change implied in a geological era.

After many years of pondering this challenge and prior to his last year of teaching at Fordham he published a seminal article called "The New Story" in 1978. In this article he observed that in the West we were between stories—biblical and scientific. This split had led to a profound disorientation of the human, alienating us from our common kinship in Earth's evolutionary processes. Seeing ourselves as apart from rather than part of Earth has led us to destroy the very sources of life in our relentless industrial assault. Without some check on this assault, Thomas Berry feared we ourselves were becoming an endangered species. In this evocative essay he was calling us back to the sources of our life, our being, our destiny in the unfolding dynamics of universe and Earth evolution. Berry's call for a New Story, a functional cosmology, was reissued ten years later in his landmark book, *The Dream of the Earth*. There he wrote: "The pathos in our own situation is that our secular society does not see the numinous quality or the deeper psychic powers associated with its own story, while the religious society rejects the story because it is presented only in its physical aspect. The remedy for this is to establish a deeper understanding of the spiritual dynamics of the universe as revealed through our own empirical insight into the mysteries of its functioning."

This is exactly what he began to do when he met cosmologist Brian Swimme in 1983 as he was working through these ideas. Brian had invited him to Chicago to give a talk at Mundelein College where he was teaching. It was an unusually cold winter in Chicago and Thomas arrived, as he often did, without an overcoat. Nonetheless, the encounter sparked warmth and excitement on both sides. Brian was so transformed by this meeting that he moved east for a year of study with Thomas at the Riverdale Center. Coming from the Pacific Northwest and having earned a doctorate in mathematical cosmology at the University of Oregon, Brian was an ideal partner for the next stage of Thomas's work. Brian's comprehensive study of evolutionary history flowed into conversation with Thomas's years of reflection on world history and religions. Thomas was delighted to be in dialogue with a scientist who understood and appreciated his perspective. Brian, too, was energized by his weekly encounters with Thomas. His horizons in human history were expanding while Thomas's empirical understanding of universe history was enlarged. With his grounded scientific knowledge, irresistible smile, and comprehensive embrace Brian became a valued participant in all the Center's events, including lecturing with Thomas in the summer conference that year.

From out of a decade-long intense collaboration, including research, lectures, and conferences, there emerged the jointly authored book, *The Universe Story* (HarperSanFrancisco, 1992). This was the first time the history of evolution was told as a story in which humans are related to the whole process and in which we are seen as having a critical role for a sustainable future. It was a fulfillment of Thomas's hope to envision the evolutionary process from the beginning as having both psychic-spiritual and material physical aspects. Human consciousness thus arises from within these processes, not as an addendum to them.

While working on this book Thomas went to an environmental conference in the Seychelle Islands in the Indian Ocean. On his return trip, flying over the Nile River, he came to the realization that he was not a theologian studying Christian theology but rather a "geologian." That is, he viewed himself as a human being who had emerged out of eons of Earth's geological and biological evolution and was now reflecting on our world. This became a context for reinventing the human at the species level, as he suggests in his thought-provoking appendix in this volume.

Thus it was after Thomas retired from teaching at the age of sixty-three that he completed his most significant writing in the area of evolutionary cosmology in relation to the ecological crisis. This included not only *The Dream of the Earth* and *The Universe Story*, but also *The Great Work* in 1999, *Evening Thoughts* in 2007, and *The Sacred Universe* in 2009. This book, *The Christian Future and the Fate of Earth*, represents a culminating moment in a long journey of struggle and reflection, a journey that spanned more than nine decades.

Conclusion

Since meeting Thomas Berry nearly forty years ago we have become more aware of the geological layers of his thinking. Over the years of working with him we realize how his thought positioned him toward broad orientations that have had organic continuity with one another. Among these layers are: the play of texts, institutions, and personalities in the history of religions; the cultural-historical settings in which religions emerge, develop, and express their deeper directions; the complex and sometimes syncretic relations between and among the world's religions; the inherent and formative relationships of local bioregions and human societies; cosmological expressions within the various religions; the awakening to our growing realization of the continuity of the human with the community of life; the evolutionary story as a functional cosmology for our multicultural planetary civilization.

That Thomas would reflect on these many layers in the context of the Christian process is noteworthy, for this comprehensive perspective becomes understandable in light of his life-long interest in and commitment to Western history, the Catholic tradition, monasticism as a contemplative way of life, and the revelatory character of science as an outgrowth of Western religious thought. His ever-broadening vision was always adventurous, as he was given to exploring at the edge of ideas that captured his imagination.

In all these reflections there remains the image of Thomas standing in his corduroy coat, talking to an audience or a class, and articulating with wonder, beauty, and creativity his dream of the Earth community fully embodied.

1.

Spiritual Traditions
and the Human Community

(1987)

In addressing the topic of the spiritual traditions and the human community, it is helpful to recall that many of the great spiritual traditions of humankind emerged out of confrontation with terror. These traditions are not the ephemeral activities of weak souls with little of that basic courage required to deal with fundamental life issues. These spiritual traditions represent humanity's ultimate confrontation with chaos, with incoherence, with destruction, with the absurd. These are not abstractions, but cosmic powers vastly different and infinitely more effective, more devastating, and more pervasive than those forces we generally think of when considering the evils to which humans are subject.

We begin to appreciate this ancient function of religion as present frustrations drive our thinking to an ever deepening level of comprehension. Behind the various manifestations of political confrontation, financial disarray, and ecological disorder lie deeper issues such as the will to self-destruction, psychic enervation, a radical incapacity for intercommunion of peoples in their group relations, and the desire to control natural processes.

We begin now to recognize the nature and dimension of the problems confronting us. We see that these are precisely the issues with which religion has been concerned from the beginning. Thus the question of humanity's future is not something to be dealt with simply in its political or economic aspects. Yet we must say that religious traditions themselves (which arose within limited cultural and social contexts) have hardly begun to function in a comprehensive human setting. Thus the responsibility for their ineffectiveness is due in large part to the religious traditions themselves, as well as to a general neglect of them by modern Western thinkers.

That this broader view of the problems we face is finding expression can be seen in two human concerns presently observed: concern for the sustaining cosmic processes of nature and concern for traditional spiritual disciplines. By virtue of the first we see that humans cannot simply do as they please with the natural world. With regard to the second we see that humans likewise cannot do as they please in relation to the psychic and spiritual forces of the world, but must also enter into communion with these powers. Our mistake has been to think that we humans have liberated ourselves from both Earthly and spiritual dimensions of our surroundings. The ideal ecological process, on the other hand, must be a complete process, one that includes the physical and the spiritual as well as the human dimensions of reality.

It could perhaps have been expected that for a period humans would investigate their own powers of scientific analysis and manipulation of nature. That study culminated in a new cosmological myth that must henceforth provide much of the basis for the future development of the human. But one phase of this experimentation is not over, and that is the test to see whether humans can control the universe in any absolute manner. Also, the testing of the deeper and darker forces of the universe is still needed, if only to learn that there are powers of a terrifying nature within the universe and within the human that can be dealt with only in union with those mysterious forces out of which the universe originated.

In antiquity, nothing was undertaken in the human order by humans alone. It had to be done in alliance with both cosmic and spiritual processes. Any integral activity involved a three-fold aspect: human, spiritual, and natural. This triple aspect was especially true in human affairs. No effective functioning was considered possible except in alliance with a broader area of reality.

The reason for the requirement is simple. The terror was there and humans were too frail to deal with it alone. In opposition to the terror were a benign providence and a beneficent universe ready to ally themselves with humans and to assure them of an inner tranquility in the larger pattern of existence, even if humans were often permitted to suffer in terrible ways and even to undergo extensive destruction. In the midst of such trials humans gained their true greatness. In the religious-spiritual context there was no ultimate defeat, for humans bore within themselves a transcendent dimension activated in its highest expression in and through such difficulties. Now, after a long period of alienation from the inner dynamics of the natural world and from communion with the higher numinous spirits, humans have awakened to these other two dimensions of reality. To some extent at least we are recovering reverence.

Attention to the natural world has culminated in global conferences held in recent years and planned for the future on energy, water, and other ecological issues. Moreover, now we perceive another area needing incorporation into our thinking. That is the realm of the sacred. In this area, also, much has happened in recent times. A pervasive sense of the spiritual dimension of reality has manifested itself on an extensive basis, not only in the United States but throughout a large part of the world as well. A certain religious initiative is now demanded of the human community. In this we can no longer say that it is the sole responsibility of religious institutions as though spirituality were separate from the human venture itself, just as we cannot say that the political venture belongs simply to a specialized group of political personalities.

There are presently four main religious issues facing the human community in its spiritual aspect. The first is the grounding of the various spiritual traditions; the second is the activation of the macrophase of each tradition; the third is the intercommunion of the traditions; and the fourth is the entry of the traditions into alliance with the newly developed cosmological myth of modern science.

As to the first of these, the grounding of the respective traditions, we are referring to a return to fundamentals. We need hardly worry about the survival of these traditions in their basic formalities, but fundamentalisms may have positive as well as less desirable aspects. An important positive aspect is that humankind needs to anchor itself firmly within primordial traditions that have a lasting contribution to make in preserving humans from that meaninglessness seeming to afflict the world at present. However, fundamentalisms that cling to intolerant positions regarding respect for other religions are surely problematic.

The second religious issue facing the human community is activation of the macrophase of each tradition, a process ongoing today. Each religious/spiritual tradition throughout the world has meaning, not only for the originating community, but for all humankind. Each tradition seems destined to be communicated on a widespread basis. This universalism we see in indigenous traditions and in Buddhism, Hinduism, Confucianism, Judaism, Christianity, and Islam, as well as in other traditions throughout the world. These are being communicated in terms more of their spiritualities than of their doctrinal expression or ritual worship and are at present powerful forces throughout the world.

The third religious issue has to do with the intercommunion of traditions. This is far advanced at the present time. It provides a context in which the deeper forces within each tradition can be activated. Therein is their challenge and their completion. Within the larger world of humankind the multiple spiritual and humanist traditions implicate each other and evoke from each other higher developments of which each is capable. Here traditions complement each other, for each has a universal mission to humankind. Each is pan-human in its significance. None can be fully itself without the others. Each has a distinctive contribution to make to human development, a contribution that is unique to each tradition. Each must, therefore, be kept distinctive even as it reaches for universal diffusion. For any tradition to withhold itself from the other cultures or for any to exclude others is to vitiate and stultify its own domain and its own development, to condemn itself to a sterile isolation from other agencies that can in these times give it life and creativity.

All human traditions are dimensions of each other. If, as Christians, we assert the Christian dimension of the entire world, we must not refuse to be a dimension of the Hindu world, of the Buddhist world, of the Islamic world. Upon this intercommunion on a planetary scale depends the future development of the human community. This is the creative task of our times, to foster the global meeting of the nations and of the world's spiritual traditions.

The global past of humanity is now the past of each human. We in the West are no longer heirs simply of the Mediterranean traditions. We are heirs of the larger human traditions, especially the spiritual elements of these traditions. These human traditions are much larger and infinitely more resplendent than the limited Western past. To deny ourselves the intimate, experiential acceptance of the large human past is to limit and cripple our present human existence. We can no longer be humans in any full sense of the word except within a global setting.

The fourth religious issue facing the human community involves identifying the spiritual dynamics of the new cosmological story given us by modern science. Of the four issues, this last is the most difficult and the most needed. Our present world situation and our future prospects have suffered from a modern scientific account of evolution poorly understood, trivialized, or basically rejected by various religious traditions. Rejected for different reasons by different traditions, the new evolutionary cosmology is refused by traditional Western religions because they are overly committed to a redemptive process and little interested in creation dynamics, the basis of the new scientific vision of reality.

The salvific, redemptive traditions of the West tend to save humans out of the temporal order or to assign meaning to the temporal order in terms of a "salvation history," with an ultimate goal outside of time. The emphasis is on trans-temporal experience. Some Hindu and Buddhist schools have also become salvific religions focused on liberation processes anchored in meditation and yogic practices. Even a religion as intimate with the cosmic process as Confucianism has at times limited itself to a seasonal cycle.

The new cosmological story, on the other hand, is a story of development, of time that needs in some manner to be validated from within. This is the story and the rhetorical context in which serious discussion of the future must take place. Merely external tolerance or even accommodation of this new cosmological world cannot be effective. What is needed is a capacity to see the spiritual dynamics of the cosmic processes within the context of developmental time. As this emerges into present consciousness it will provide an integrative context. The relentless developmental mode of cosmic unfolding is reaching a new level of integrative, reflexive understanding.

Within this context a further story proper to the fourth phase is being born. This integrative story is finding its proper function in the new ecological consciousness that is spreading throughout the world. Religiously this stage is finding its expression on a global scale of spiritual awakening. Nothing is likely to force the new development so powerfully as an awareness of the destructive forces threatening to erupt throughout the world if we do not succeed in this effort.

The deepest energies of all human traditions are needed to deal with contemporary challenges. The most profound human disciplines are needed, for the final decisions in economics and politics as well as in military and diplomatic affairs are ultimately made, not by technology or computers, but by humans subject to both visionary hopes and moods of desperation. But most of all at the present time there is a need to work with sustained human energies through periods of great darkness.

That we will succeed is to some extent assured by the whole of past history, by the forces that sent the galaxies into space, that shaped Earth and brought forth living creatures in the sea, on Earth, and in the sky. It is to some extent also assured by the course of human history with all of its agonies and catastrophic periods. We cannot expect to achieve anything substantial without upheaval. Its full dimensions we do not know. The future is also hopeful because of the integral, cosmological story available to us now for the first time.

2.

The Third Mediation

(1982)

Each age of human history has its distinctive work to accomplish. Israel in the prophetic period had as its special task to present humankind with a sustaining vision of future historical transformation. The early Christian ages were assigned the task not only of resisting the tyranny of the Roman imperium but also of reconciling the Christian message of redemption with the great philosophical traditions of the Greek and Roman worlds. Christians in the medieval period had their special task in raising up a new civilization out of the ruins of the Roman period and the tumultuous movements of diverse ethnic peoples across Eurasia. So, now, we ourselves have a special task to fulfill, the task of the "Third Mediation."

The *First Mediation* is that between the divine and the human, a mediation begun in ancient Israel, continued in the redemption accomplished by Christ, and communicated to Mohammad in later centuries. This process has dominated most of the last three thousand years of human history in the West.

But, while this concern with reconciliation between the divine and the human has continued as a central preoccupation of the Christian tradition, we have found it necessary during the past two centuries to give special consideration to the *Second Mediation,* the inter-human mediation, the reconciliation of differing human groups. With the rise of the industrial establishment these groups divided into classes antagonistic to each other. The period, moreover, saw the rise of the great nation-states, each so absolute in its demands that it could not tolerate opposition or injury from any other state. Thus emerged the major international and world conflicts that have caused such extraordinary destructive power over the past two centuries. Thus, too, evolved the great social and revolutionary movements that have shaken our world.

Dealing with these powerful conflicts has required special attention to inter-human arbitration as a distinctive aspect of our human and Christian task in recent times. This mediation has become ever more urgent in its demands as the human community now spends over a billion dollars each day—four hundred billion dollars each year—for military purposes.

Our preoccupation with this second mediation continues even while a *Third Mediation* has become an imperative so strong that it overshadows in its significance even its predecessor. I speak of the mediation between the human community and Earth, the planet that surrounds and supports us and upon which we depend in an absolute fashion for our nourishment and our breath. In speaking of Earth, I include, too, the sunlight by which all things live, by which we behold the visible things about us, and by which we have the warmth needed for survival. Even this sunlight, coming to us from such a great distance, is communicated to us by Earth.

It is not only food for the body that comes from Earth, but our very powers of thinking and the great images in our imagination. Our arts and education, too, all proceed from Earth. Even our knowledge of God comes to us from our acquaintance with Earth, for the divine reveals itself first of all in the sky and in the waters and in the wind, in the mountains and valleys, in the birds of the air and in all those living forms that flower and move over the surface of the planet.

Yet, despite this fundamental dependence on Earth, we have during these past two centuries severely damaged our planet with massive technological strategies and machines. That these inventions do much good is surely true; that they have a demonic aspect is also true. As we bring about the poisoning of air, soil, sea, and all living beings inhabiting these regions, we begin to make of Earth a place where the existence of higher life forms is threatened on a planetary scale. A forceful presentation needs to be made to prevent the destruction of the divine image as presented to us in the created world, to prevent the wiping out of the sources of our spiritual, intellectual, and aesthetic development, to prevent the burning away of irreplaceable resources for foolish and ephemeral purposes. Unless we are totally depraved, we will seek to give to our children not only life and education but a planet with pure air and bright waters and fruitful fields, a planet that can be lived on with grace and beauty and at least a touch of human and earthly tenderness.

So far Christians have not distinguished themselves by their concern for the destiny of Earth. Now, however, this care has become the special role, not only of Christians, but of all humankind, a role no other age could fulfill, a role so important that there may not *be* another truly human age in the future if the present conflict of humans with Earth is not resolved, if this role of the third mediation is not fulfilled. The ultimate danger of war in the future is not only that it will destroy a multitude of human beings, but that it might also render Earth itself inhospitable to higher life forms. Only the lower forms of the vertebrates and swarming communities of insects might then possess planet Earth.

When we turn to examine the resources that Christians possess for fulfilling the task of the third mediation, I would mention first the vast Christian community throughout the world. While Christians have often been estranged from each other in the past, we now begin to reflect on the basic unities binding us together as a people. To see this third mediation as a preeminent Christian task in our times is to begin formation of a powerful planetary force that will hopefully become effective on the scale needed. But even while we speak here of Christian peoples, I would also mention the union of all religious peoples throughout the world who are now concerned with this task of the third mediation. This task is not simply economic or political—it is preeminently a religious and spiritual task, perhaps the most urgent task of all. Only religious forces can move human consciousness at the depth needed. Only religious forces can sustain the effort that will be required over the long period of time during which adjustment must be made. Only religion can measure the magnitude of what we are about.

(2) Our second spiritual resource is our Christian awareness of the communion to which we are called: communion with the divine source of our being, communion with the entire human community, and, finally, communion with the universe itself. While we have recognized the inseparable nature of communion of God with the human community, we have not yet realized that this communion, to be perfect, must include communion with Earth. This is the unique awareness that begins to take place in our times. The Body of Christ is ultimately the entire universe. Otherwise, neither the incarnation nor the redemption is complete. Experience of this communion is so strengthening, so ecstatic, that it can provide the energies needed to carry life on into the difficult future.

Christianity, as well as most other great religions, has been excessively oriented toward transcendence. A true Earth consciousness needs to be developed. Further, not only has divine transcendence been an overwhelming preoccupation, but human transcendence of the natural world has been also emphasized. Now we need a greater sense of humans, not as transcending the Earth community, but as members of the Earth community. If God has desired to become a member of that community, humans themselves should be willing to accept their status as members of that same Earth community.

(3) Our third spiritual resource is awareness of the creative possibilities of chaos and destruction. That order and beauty and creativity are intimately related to disorder, turmoil, and confusion can be seen in the biblical story of creation, in the experience of the prophets, in the redemptive sacrifice of Christ, in the emergence of the medieval Christian world out of

the dark centuries following the decline of Rome. The fact that our times are so destructive forces us to a level of thought, reflection, and spiritual renewal such as we might never otherwise attain. It might also presage the shaping of a world more resplendent than we have ever known before.

(4) Our fourth spiritual resource is the sustaining energy available to us in our Christian forms of worship, in our rituals, and in our sacraments. These expressions of faith are especially important, for religion is not a sentimental feeling or a pious attitude or an escape from the real challenge of life. It is, rather, a way of dealing with the hard and difficult and threatening moments of life and with the terrors of death. No other force has yet been discovered by humans that can so sustain us in adversity, so inspire us in moments of exaltation, or so awaken our imaginative and creative powers. Nor is religion a stiff, unyielding fixation made for some other age, or for some other place, or for some other issues different from those we face at the present time. Religion is, rather, as adaptable as life itself, as real as the Earth we stand on, as present as the air we breathe. Christian faith, particularly, includes not only faith in God, but faith in the human community, and faith in the Earth community.

Religion rectifies not by domination but by invocation. This is the attitude and the power needed. Indeed, our difficulties have been caused principally by a certain distrust of Earth and by a managerial mania seeking to replace or manipulate the marvelous variety and interlacing of the interior life forces of nature with mechanistic processes and chemical concoctions ultimately ruinous to the entire biosphere, the great web of life encircled by Earth. A new appreciation of and confidence in Earth is needed, along with a capacity for communion with it. Only through this comprehensiveness can we really have community. Only through an integral community can we survive.

In conclusion, I would suggest that the fulfillment of this third mediation, the establishment of a harmonious relation between humans and Earth, may well be the origin of a more effective inter-human mediation as well as a more fulfilled divine-human one. I might note, finally, that our Lenten season is not merely a religious or human ritual. It has its origin in a profound Earth renewal process. The redemptive sacrifice of Christ was associated with the springtime renewal of Earth. That same sacrifice enables us to join in this great renewal event. We do not, then, proceed with our human tasks simply by our own individual efforts. We are sustained by the powers of heaven and Earth present to us in our eucharistic celebration. A new stage of interaction between the divine, the human, and the natural is begun. With dedication we will carry it through to completion.

3.

The Catholic Church
and the Religions of the World

(1985)

Presently, Catholics constitute some 16 percent of the human community. Catholics no longer exist in a completely unified religious society as in the European medieval period. The once Catholic world of Europe has broken up into numerous Christian groups. Also, because of immigration and conversion, the Western world has experienced a proliferation of individuals and groups identifying themselves with non-Christian religions.

Abroad, the Church—which numbers over one billion people—is scattered throughout the continents and peoples of the world. The Church has grown significantly in Africa and Latin America. In addition, many large and small Catholic communities exist in authoritarian nations dominated by a militant secularism. A more tolerant secularism pervades the rest of the world and tends to dominate life ideals, as well as national and international institutions. In view of all this diversity, the religious situation of the Church and of the Catholic community has become exceedingly complex. The need for a more formal consideration of the Church in its relation to the religious and non-religious world about us is clear.

The second Vatican Council (1962–1965) gave much attention to the subject in several of its documents, notably those treating the Church in the modern world, ecumenism, missions, and the Church in its relation to non-Christian religions. We are concerned here with this last document, *Nostra Aetate* (In Our Age). It is a brief, incidental declaration of the Council, which evidently felt itself forced to say something significant without venturing very far into a subject too sensitive for any thorough treatment at the time.

What was said with extreme caution is, however, worthy of serious consideration, since now, more than ever, the parish priest as well as any teacher of religion is constantly faced with questions concerning other religious traditions. Many of our best people, especially our young people with their strong religious attractions, often find the spiritual disciplines and meditational practices of Asian religions or of the Sufi orders of Islam more helpful than what is available to them in our Church preaching or even in our religious communities. All of these non-Christian traditions, including those of Native Americans, are undergoing a vigorous phase of development in what might be considered a type of renewal. This revitalizing includes intellectual self-understanding, spiritual practice, and social presence. Both Asian traditions and Christian movements experiencing renewal are in reaction to the trivialization resulting from so many efforts at adaptation to the modern secular world.

The effort at traditional integrity and renewal in various Christian religions exists along with increased study of the influence of a particular tradition on other religions, as well as their influence on it, especially the influence of those religions considered non-Christian. In this respect the Christian West has probably been more influenced by the other religions than those traditions have been influenced by the Christian West. Both have been overwhelmingly influenced by modern secularism.

In this situation the Church has sought guidance for its actions from its own biblical, patristic, and theological teachings. The Second Vatican Council's 1965 declaration *Nostra Aetate*, for example, refers to biblical passages asserting the divine concern for all people. In a rather guarded manner the statement is made that other religions often reflect "a ray of that truth which enlightens all men." Immediately following this comment is a reference to Christ as "the way, the truth, and the life, in whom men find the fullness of religious life, and in whom God has reconciled all things to Himself" (1 Corinthians 5:18–19). The quotation betrays a certain anxiety lest admission of any authentic revelatory experience outside the Christian tradition might lead to a diminution of the Christian claim to the integral revelation of the divine to the human community.

This way of dealing with the inter-religious issue has its own validity, though it is not particularly new in its overall vision. Already, throughout past centuries, substantially the same thing has been said, although now religious studies identify in greater detail ways in which these other religions reflect, not only "a ray" of the divine light, but even floods of light illumining the entire religious life of the human community. The Council did clarify in a few succinct statements the Catholic view that revelation is not absent from the non-Christian world; that all that is spiritually good should be encouraged and developed in Native traditions; and that the peoples of Earth comprise a single community with a single origin and a single destiny, and that a single Providence communicates its saving design to all peoples.

Behind these statements are the basic questions of unity and diversity and how differences relate to each other in some meaningful unity. In the attitude of the Council, and in the Christian tradition from the beginning, we find a powerful sense of unity accompanied by a suspicion if not abhorrence of diversity in religious concerns. This attitude originates in an overwhelming sense of the oneness of a personal, transcendent, divine creator. Originally experienced as a tribal deity, Yahweh absorbs into himself all that divine power generally experienced as diffused throughout the universe and articulated on planet Earth in the manifold phenomena of the natural world. Associated with this deity is a singular people bound by a covenant expressed in written form, based on which a later new covenant expression is communicated to the peoples of Earth through a divinely established hierarchical Church.

This sense of an elect people as exclusive bearers of a universal salvation either originated in or was powerfully reinforced by the feeling of a small people who had an important destiny, a people constantly threatened by surrounding political powers. The more threatened this elect people felt, the more intensely they experienced their own significance as a people ordained to be the instrument of divine rule over all the nations of Earth.

This attraction to an ambiguous political-religious rule by an elect people was spiritualized in the New Covenant period. The Church did emerge with the sense of having an exclusive universal role in bringing about the spiritual well-being of a fallen world. Other religions of the world were seen fundamentally as obstacles, although the Holy Spirit never failed to be present to well-disposed individuals. While a multitude of scriptural, patristic, and theological quotes can be offered in support of a larger overall vision of the salvific process, these are generally extenuations or remedies for a fundamentally undesirable situation.

Moreover, the unity of religions has often been viewed in the Catholic Church from an exclusivist, triumphalist perspective. Because of this until recently there has been a lack of appreciation of that diversity which Saint Thomas designates as the "perfection of the universe." By definition the universe is diversification caught up in a complex of functional relationships. Obviously the diverse part in its particularity cannot itself constitute the principle of unity, although each part articulates the whole in some unique fashion. The perfection, however, is in the whole, not in the part as such. Here it would be of some help to quote from that same article of Thomas Aquinas in the *Summa Theologica*. Dealing with the distinction of things, Thomas remarks in a concluding statement (in I, q. 47, a. 1) that the multitude of things comes from the first agent who is God:

> For he brought things into being in order that his goodness might be communicated to creatures, and be represented by them; and because his goodness could not be adequately represented by one creature alone, he produced many and diverse creatures, that what was wanting to one in the representation of the divine goodness might be supplied by another. For goodness, which in God is simple and uniform, in creatures is manifold and

divided; and hence the whole universe together participates in the divine goodness more perfectly and represents it better than any single creature whatsoever.

This law of diversity holds, not only for the other areas of
being and of action, but also for the religious life of the human
community, for revelation, belief, spiritual disciplines, and
sacramental forms. If there is revelation, it will not be singular
but differentiated. If there is grace, it will be differentiated in its
expression. If there are spiritual disciplines or sacraments or
sacred communities, they will be differentiated. The greater the
differentiation the greater the perfection of the whole, since perfection is in the interacting of diversity; the extent of the diversity is the measure of the perfection.

When the religious traditions are seen in their relations to
each other, the full tapestry of the revelatory experience can be
observed. Each articulated experience is shared by the others. It
is in the fabric of the whole that the divine reveals itself most
fully. This fullness requires a threefold sequence of emphases in
the various traditions. First, there is the primordial experience
expressed in an oral or written form, the scriptural period. This
occurs by means of an isolation process in which the tradition
draws away from the surrounding ethos and, in doing so,
secures its identity. Second, there is the deepening of the tradition, a patristic period during which the implications of the original experience are elaborated by individuals in contact with the
larger life process. Third, there is the period of expansion, of
interaction with other traditions in their more evolved states.
These three phases are not mutually exclusive, since some interaction of traditions is present from the beginning, both in assist-
ing positively and in providing a polarity of opposition.

The importance of the isolation and interior development
phases can be seen in each of the major traditions that have so
powerfully influenced the religious life of the human community.
In India, for instance, the profound mystical and metaphysical
developments of the early first millennium BCE required a special
type of psychic intensity. In this sense the Upanishadic wisdom
articulated in the Brahman-atman (Absolute-self) identity both

draw on earlier Vedic cosmological thought and draw away from a pantheon of personalized deities. In China, the focus of attention was much more cosmological. The divine was understood as Shang-ti (High God), or as T'ien (Heaven), as the great mystery presenting itself in the vast cosmological cycles in which the human was also a functional presence. The divine, the natural, and the human were thus present to each other in the grand sacrificial reality of the universe itself and in its rhythmic pulsations. This cosmic sense is also present for the Japanese with their aesthetic expression of the numinous in the natural world. Their cultivation of spiritual simplicity and spontaneity in relation to nature is unique in the human community. So too is the presence of an ultimate creative mystery throughout the natural world and in the human heart found in Native American traditions. In every case, these ultimate orientations toward reality and value originate in an interior depth so awesome that the experience is perceived as coming from a trans-phenomenal source, as being revelatory of the ultimate mystery whence all things emerge into being.

To maintain that these experiences are simply the consequences of natural reason and not valid revelatory experiences communicated by the divine would be to negate the sacred character of these most profound of all experiences shaping the human psychic structure. Certainly India is absolutely clear on the need for the divine to reveal itself in some active, positive manner if humans are to know the divine in any effective manner.

Many of the best Asian traditions indeed have explicit statements on the need for grace, for a special mode of divine activity to effect the spiritual fulfillment of the human. Through the support of these sacred experiences not only Asians but people everywhere have been able to attain sublime spiritual insight and to endure and even to exult in a life of much struggle and pain, along with ecstatic delight. The language, customs, and artistic styles of all humans give expression to this presence of the divine. Each of these revelatory traditions in its own mode reaches unequaled levels of religious experience.

All of these traditions were substantially complete in their earlier expression. Hinduism in its proper line of development will not likely go beyond its expression in the Vedic hymns, the

Upanishads, the epics, the *Bhagavad Gita*. Buddhism in its proper line of development will hardly go beyond its expression in the early dialogues, the *Dhammapada,* the *Sutta Nipata*, the verses of Nagarjuna, the *Lotus Sutra*, and the *Vimalakirti Sutra*. This is also true for the Four Books of the Confucian classics, the *Tao Te Ching*, and the writings of Chuang-tzu. These writings are full and perfect in their own context, although each expresses a vital expanding process with an ever-renewing series of transformations through the centuries.

So, too, the Jewish, Christian, and Muslim revelatory experiences are full and complete in their own proper mode. Judaism, Christianity, and Islam experience revelation through a sequence of historical events as opposed to the metaphysical, the mystical, the cosmological, the psychological, or the immediate experience of the numinous dimension of the natural world. While all these traditions developed in limited geographical areas and modes of consciousness, each revelatory experience was considered a comprehensive interpretation of the universe and an effective guide for individuals and communities in attaining their divinely determined destiny. Although this sense of completeness existed in the various traditions, they have generally been open traditions, willing to interact creatively with other traditions through both listening and speaking.

The deepest values of each tradition, however, have been in their own distinctive insights. It would not have been possible for India's experience of the trans-phenomenal world to be developed in its full intensity simultaneously with the biblical experience of the revelatory aspect of historical events. Nor could China have developed its insight into the mysterious Tao of the physical universe simultaneously with the high metaphysics of India. Nor would the experience of the Great Spirit manifested in the natural world by the indigenous peoples of North America be compatible with a simultaneous experience of the Buddhist doctrine of emptiness or the Neo-Platonic doctrine of the Logos.

None of these experiences rivals the others. Each needs a certain isolation from the others for its own inner development. Each is supreme in its own order. Each is destined for universal diffusion throughout the human community. Each is needed by the others to constitute the perfection of the revelatory experience. Furthermore, each carries the whole within itself, since part and whole have a reciprocal relationship. Each has its microphase-initiated institutional adherents and its macrophase or universal presence throughout the human community. We are now living in the macrophase period of development of most religious traditions, the period of extensive influence without formal initiation. This situation is possible because none of the traditions does precisely what the others do. Each has, as it were, its own area of spiritual consciousness. The traditions are thus dimensions of each other.

In this perspective, the Second Vatican Council document on revelation should be entitled "Christian Revelation" or "Biblical Revelation," since it does not deal with the revelatory experience in other religions. Also, any reference to Christian revelation as the "fullness" of revelation must consider the precise sense in which this term is being used, since the Christian scriptures and the Christian tradition themselves indicate that there exists a revelatory communication of the divine throughout the human community. If this idea of revelation be true, then the fullness of the revelatory experience is in the larger range of these highly differentiated experiences. Each tradition carries the fullness of revelation implicitly in itself.

The difficulty in statements about the Holy Spirit being present to all peoples is the implication that the Holy Spirit is communicating to others the same thing revealed in the biblical-Christian tradition, only less clearly or less completely. The difference is seen as quantitative rather than qualitative. If the difference is qualitative, then, according to traditional views, these traditions do not qualify as "revelation" but as some "natural" mode of knowing. If the difference is quantitative, then, according to traditional views, Christian superiority can be stated in terms of fullness or completeness.

The first statement is where the difficulty might be located. Within the authentic revelatory process itself, I am proposing a qualitative difference, a difference that cannot be resolved in terms of fullness or completeness, but only by the mutual presence of highly differentiated traditions. What is to be avoided is any uniform tendency in the meeting of religions. What is to be sought is a mutually enhancing meeting of qualitatively differentiated religions in which both the microphase and the macrophase expression of each religion would benefit.

The expression "one flock and one shepherd," if not understood properly, is a seductive attraction that easily leads to sterility in the religious process. It suggests an attitude that diversity should not be, that it is a hindrance to human well-being and to the salvific process, and that it must be ended as quickly as possible. This outlook has led to such an aversion toward other religions that nowhere on the North American continent is there a Catholic university where graduate studies on other religions can be done adequately. Nor, as of the last years of the twentieth century, is there concern for these other religions in our theological seminaries. The History of Religions program that I established at Fordham University, begun in 1968, was discontinued in 1980.

Because of this misconstrued ideal of "one flock and one shepherd," Catholic interaction with other religions is diminished in its efficacy. Since the basic principle of life and movement is unbalance rather than balance, asymmetry rather than symmetry, the challenge is avoided and the tension needed for creativity is lacking. Diversity is enrichment. For the biblical concept of deity to be the universal concept to the elimination of *Shiva* and *Vishnu*, of *Kuan-yin* and *Amida*, of *Shang-ti* and *T'ien*, of *Orenda*, *Wakan-tanka* and the *Manitou*, would be to impoverish the concept of deity. To consider the Christian Bible to be the only scripture, to the elimination of the Vedic hymns, the *Upanishads*, and the *Bhagavad Gita* of India; to the elimination of the *Qu'ran* of Islam, the *Lotus Sutra* of Buddhism, the sacred Books of China, would be to constrict rather than to expand our understanding of divine-human communication.

For any situation the ideal is the greatest tension that the situation can bear creatively. Although every archetypal model needs multiple realizations, the sacred, more than any other element of reality, needs variety in its modes of expression. The difficulties experienced by Christians in accepting the variety of religious traditions can be resolved:

1. By distinguishing between the microphase membership and macrophase influence of all religious traditions

2. By identifying the unique communication of Christian revelation in both modes of its expression (the natural world and scripture)

3. By recognizing the qualitative differences in religions and fostering these differences

4. By identifying the creative dynamics of inter-religious relations

5. By fostering a sense of the New Story of the universe as context for understanding the diversity and unity of religions

Our discussion so far has been within the general context of traditional modes of thinking, with no significant reference to the vast changes in our modern way of experiencing the universe, the human community, and the modes of human consciousness. Yet these are powerful determinants in all our religious as well as in all our cultural developments.

To talk about religious traditions simply out of their own inner processes is to ignore this larger context of interpretation. Accepting this new context is difficult for us because of the prevailing secular culture, the materialist view of the universe, and rationalistic modes of contemporary thinking. These attitudes have been so antagonistic to religion that we can hardly believe the long course of scientific meditation on the universe has finally established the emergent universe itself as a spiritual as well as a physical process and the context for a new mode of religious understanding. We might describe it as a meta-religious context for a comprehensive view of the entire complex of religions.

We are rewriting *The City of God* of Saint Augustine, not this time as the story of two cities seeking each other's extinction but as the story of an immense cosmic process, both spiritual and physical from the beginning, articulating itself in ever greater variety and complexity until it has come to a certain fullness of expression in human consciousness. Within human intelligence the creative process attains a capacity for self-awareness and for a human inter-communion with the numinous mystery present throughout this process. This inter-communion, as a revelatory presence of the divine, takes place throughout the human community in the diversity of its manifestations. From these primordial indigenous experiences have come the diverse scriptures of the world, the various forms of worship, and the variety of spiritual disciplines.

I suggest this context of interpretation for the diversity, identity, and inter-communication of religions. It might be considered as a cosmological-historical approach over against the traditional theological, sociological, or psychological approaches to the subject. This cosmological approach accords with the basic statement of Saint Thomas concerning the cosmic community as the "perfection of the universe," as the supreme reality which "participates in the divine goodness more perfectly, and represents it better than any single creature whatsoever." It also accords with the view of Pierre Teilhard de Chardin (1881–1955) that "[The human] is a *cosmic* phenomenon, not *primarily* an aesthetic, moral, or religious one."[1]

If the human can be understood only within the universe process, then every aspect of the human, including the religious dimension, is involved. But whether we begin with Augustine, with Thomas Aquinas, or with Teilhard, the universe is the primary religious reality, although this religious dimension, as well as its psychic dimension, is fully articulated at this time only in the human. In no instance, however, is the human in any of our activities functioning simply by itself, independent or isolated from the universe process that brought us into being, sustains, enlightens, and sanctifies us.

If humans have learned anything about the divine, the natural, or the human, it is through the instruction received from the universe around us. Any human activity must be seen primarily as an activity of the universe and only secondarily as an activity

---

[1]*Letters from My Friend, Teilhard de Chardin,* ed. Pierre Leroy (New York: Paulist Press, 1976), 137; italics added.

of the individual. In this manner it is clear that the universe as such is the primary religious reality, the primary sacred community, the primary revelation of the divine, the primary subject of incarnation, the primary unit of redemption, the primary referent in any discussion of reality or of value. For the first time the entire human community has, in this story, a single creation or origin myth. Although it is known by scientific observation, this story also functions as myth. In a special manner it is the overarching context for any movement toward one creative interaction of peoples or cultures or religions. For the first time we can tell the universe story, the Earth story, the human story, the religion story, the Christian story, and the Church story as a single comprehensive narrative.

The choice, however, of Vatican Council II was to establish the biblical redemption story rather than the modern creation story as its context of understanding. In doing so, it set aside its own most powerful instrument for dealing with the Church, revelation, the modern world, missions, and relations with non-Christian religions. It will undoubtedly be a long time before such a transition in our thinking will take place. We might propose, however, that until this new context for understanding is accepted, the unique role of Christianity and of the Catholic Church will not attain its full efficacy. If Saint John and Saint Paul could think of the Christ form of the universe; if Augustine could realize that the beauty that he came late to love was in the rhythm of life; if Saint Thomas could say that the whole universe together participates in the divine goodness more perfectly and represents it better than any single creature whatever; and if Teilhard could insist that the human gives to the entire cosmos its sublime mode of being, then it should not be difficult for us to engage the universe itself as the primordial sacred community.

4.

Christian Cosmology

(1985)

The need of a new context of interpretation for presenting the Christian message was indicated by Bishop Malone, president of the National Conference of Catholic Bishops, in his Report to the Synod Secretariat, where he says quite directly: "The Church stands in need of a new symbolic and affective system through which to proclaim the Gospel to the modern world."[1]

What we propose here is that the need is not precisely for a new "philosophy" in the general meaning of that term, but for a new "cosmology." Our world is not presently controlled as such by its philosophy but rather by its cosmology. Our modern cosmology is associated with the observational sciences, whereby we have come to understand our universe in terms of its evolutionary unfolding over vast periods of time and through the vast extent of space. Presently, Christians are not thinking or acting within this accepted story of the universe. Yet outside of this story we can do very little in making the Christian vision visible.

Just as Clement of Alexandria presented Christianity within the context of Hellenism, as Augustine worked within the Neo-Platonic frame of reference, Thomas within the Aristotelian

---

[1]*Origins* (September 26, 1985): 233.

perspective, and Ignatius within the context of Renaissance humanism, so now there is need for us to think and act and create within a new period of Western Christianity by working within the context of this New Story of the universe. We are presently quite comfortable religiously, theologically, and even ministerially within the walls of our own seminaries, churches, and religious houses. Outside of this setting however, we work with minimal efficiency.

We are assuredly not providing the kind of exciting leadership needed for these times. That insight is sufficiently evidenced by the fact of our declining priesthood and religious communities. The difficulty is not that youth are less motivated or less competent, but that the more competent and spiritually motivated cannot find within our establishments any adequate context in which to work effectively on the greater issues presently confronting the human community. They wish to work not simply for redemption out of the world but for the survival and enhancement of life on a planet profoundly threatened in all its basic life systems. What they see in the Church is an establishment taking no serious notice of this situation. Our church authorities, universities, theologians, and Catholic media seem to be showing no significant interest in the fate of Earth as it is being devastated by a plundering industrial system.

As Christians, the question of human-Earth relations seems outside our concern. So overwhelming are our religious and social concerns that we fail to recognize that both our social and our religious well-being largely depend on the well-being of Earth, which provides sustenance for our physical, imaginative, and emotional as well as our religious well-being. To understand planet Earth and our intimate relationship with it, we need to know the great story of Earth and of the universe that brought it and ourselves into being.

What needs to be recognized is that this New Story of the universe represents the greatest change in human thought and consciousness since the rise of the Neolithic Period. Thus it is not only a difficulty for Christians, it is a difficulty for human consciousness throughout the Earth community. One of the reasons for this problem is that the New Story of the universe has brought with it such enormous powers that we are presently engaged not simply in historical or cultural change; we are changing the chemistry of the planet, its biological systems, and even its geological structure. In each of these areas, the human presence in the twentieth century has brought about a profoundly disturbed situation. The human has become not the crowning glory of Earth, but its most destructive presence.

The solution is not, then, a case of simply restoring a former religious, spiritual, moral, or humanist tradition. It is a case of re-ordering the human in its entire relationship with the planet on which we live, a mission for which Christians are not especially suited, since we have seldom shown any extensive regard for the creation process, dedicated as we have been since the thirteenth century to a primarily redemptive task. Both the Apostles Creed and the Nicene Creed glide lightly over any reference to creation. So, too, does the Catechism of the Council of Trent and the Baltimore Catechism that was derived from it. None of these shows any significant concern for the created world, or the "natural" world as this is often designated to distinguish it from the "supernatural" world, which is our primary concern. In our prayers at Mass we constantly pray for deliverance into a "better" world. Because we have in the past brought our Christian beliefs in contact with various "philosophies," we still think of turning to a new "philosophical and conceptual framework" for the remedy of our present difficulties and the rendering of our work more effective in the modern world. The real need was indicated over fifty years ago by Pierre Teilhard de Chardin (1881–1955) when, after journeying across the North American continent, he wrote the essay "The Spirit of the Earth" in his treatise on *Human Energy*. This is not exactly a "philosophy," but an interpretation of religion and spirituality in terms of the story of the universe and of the Earth process.

It is to communicate something of this story that the following pages are written as a suggested outline of the paradigm for understanding what is available to us at the present time. We might even claim that this view of the universe provides a more comprehensive context for interpreting Christian belief than do the views of Clement, Augustine, Thomas, or Ignatius.

The new paradigm for the understanding of reality and of value can be identified in the following manner. *First,* the new paradigm sees the universe as a sequence of irreversible transformations begun some 13.7 billion years ago. This story of the universe as told in its four major phases—as the galactic story, the Earth story, the life story, and the human story—constitutes the new pattern or paradigm of intelligibility. The entire scientific venture of the past few centuries culminates in a capacity to tell this story with amazing insight into the sequence of events and their interdependence from the beginning until the present. This story provides the basis for our sense of reality and of value.

Second, the evolutionary process of the universe has from the beginning a psychic-spiritual as well as a material-physical aspect. There is no moment of transition from the material to the psychic or the spiritual. The sequence of development is the progressive articulation of the more spiritual and numinous aspects of the process. If, for a period, this story was told simply in its physical aspect to the neglect of its psychic aspect, this is no longer adequate. The period of preoccupation with quantitative material processes seems to have been necessary for penetrating the deeper structures and functioning of the universe. But the unfolding of the universe from lesser to greater complexity and consciousness is now widely understood.

Third, Earth has a privileged role as the planet whereon life is born with all those special characteristics we find in Earth's living forms. The unity of the Earth process is especially clear. It is bound together in such a way that every geological, biological, and human component of the Earth community is intimately present to every other component of that community. Whatever happens to any member affects every other member of the community. Here we can see how precious Earth is as the only living planet that we know, how profoundly it reveals mysteries of the divine, how carefully it should be tended, how great an evil it is to damage its basic life systems, to ruin its beauty, to plunder its resources. For these things to be done by Christians or without significant Christian protest is a scandal of the primary order of magnitude.

Fourth, the human is by definition that being in whom the universe reflects on and celebrates itself in conscious self-awareness. The human is a mode of being of the universe entire as well as a distinctive being in it. More than any other being the human is intimate with the universe in the full range of its extension in time and space. Teilhard expresses this thought in a letter written in 1952: "Man can only be understood through tracing his rise through physics, chemistry, biology, and geology. In other words, man is a cosmic phenomenon, not primarily an aesthetic, moral, or religious one."[2]

This description gives humans their proper dimensions and their proper role in the universe process. Only in such a comprehensive perspective can we understand how, through its human expression, the universe returns to itself and to its numinous origins. In fulfilling this role, humans depend on the natural world for every aspect of their intellectual insight, spiritual development, imaginative creativity, and emotional sensitivity.

This cosmic dimension of the human has consistently been recognized by the various peoples of the world. Although it is the basis for the cosmic dimension of Christ recognized by Saint John and Saint Paul, it has not in recent centuries received its proper development in Christian thought. Now, however, even scientists recognize the "anthropic" principle of the universe. As physicist Freeman Dyson tells us: "The more I examine the universe and study the details of its architecture, the more evidence I find that the universe in some sense must have known we were coming."[3] Likewise physicist John Archibald Wheeler, after studying the conditions for producing life and consciousness,

[2]*Letters from My Friend, Teilhard de Chardin*, ed. Pierre Leroy (New York: Paulist Press, 1976), 137.

[3]Freeman Dyson, *Disturbing the Universe* (New York: Basic Books, 1979), 250.

concludes: "Why then is the universe as big as it is? Because we are here!"[4]

Fifth, the basic referent in terms of reality and of value is the universe in its full expression in space and time. The ultimate created value cannot be found in any one part of the universe, but only in the collective whole, as Thomas Aquinas says. While Thomas was thinking in a spatial rather than in a time-developmental mode of consciousness, his principle of the greater value of the more comprehensive reality remains valid for the sequential processes whereby the universe and planet Earth articulate themselves. In virtue of this principle we need to avoid that anthropocentrism that would make the human in itself so absolutely the norm of value that we fail to recognize that the larger concern must be for a universe that includes the human, but is a greater reality than the human. The lack of appreciation of this principle has led to a terrible devastation of the planet on the supposition that this devastation was simply a price that had to be paid for the exaltation of the human.

Ultimately the well-being of Earth and the well-being of the human must coincide, since a disturbed planet is not conducive to human well-being in any of its concerns—spiritual, economic, emotional, or cultural. A new referent in terms of our moral activity is indicated here. If formerly the moral norm was simply that a being should act according to its fixed nature, this norm must undergo extensive reconsideration in a world of developmental process where natures are not fixed but in process. A new and most significant norm of reference concerns that which enables the larger creative process to continue its proper line of development.

Sixth, the universe itself can be understood as the primary revelation of the divine. The numinous presence of the divine throughout the natural world has always been recognized by the various peoples of the world. Saint Paul tells us in the first

---

[4] John A. Wheeler, "The Universe as a Home for Man," *American Scientist* 62 (November–December 1974): 133.

chapter of his Epistle to the Romans that from the beginning of the world "we came to know the invisible nature" of God through "the things that have been made." More recently the verbal revelatory experience and its written form have taken such precedence in our thinking that we have become alienated from the revelatory import of the natural world.

For these reasons we have little awareness that if a resplendent world gives us an exalted idea of God, a degraded world gives us a degraded idea of God. Already this alienation from the revelatory presence of the divine in the surrounding universe is making the entire religious life more of an artificial construct. Particularly at the present time when our society has lost vital contact with its written scriptures and the traditional forms of religious expression, this experience of the revelatory import of the universe itself takes on special value.

Seventh, biblical revelation, the incarnation, redemption, and the shaping of the Christian community—all these have taken place within the larger cosmological and historical context. We can no longer understand the Bible, the incarnation, redemption, or the mission of the Christian community simply in the context of past explanations. The new mode of perception provides a more expansive context for presenting all the basic Christian mysteries.

Biblical revelation through its perception of the divine in the unfolding of history provided a historical context for the movement that eventually brought us to our present perception of the universe itself as evolutionary process. Even though Christians generally have had difficulty accepting the evolutionary story of the universe, Christianity has a basic compatibility with this account of the universe as unfolding process. Without the perception of meaning in the irreversible unfolding of historical events we might never have come to such an understanding of the universe. We can, then, look upon this larger story as the fulfillment of the biblical narrative itself. The period of creation has not been closed; it is still in progress. We are still in the sixth day of creation. In and through ourselves the world is coming into being. One of the main clarifications of creation given by the New Story is the genetic relatedness of every being in the universe with every other being in the universe. Every living being is cousin to every other being.

Incarnation: By our perception of the human as a mode of being of the entire universe we have an appropriate context for understanding the cosmic dimension of the Christ reality. The universe from its beginning had a Christ dimension in accord with the statement of Saint John concerning the Logos that "all things were made through him and without him was not anything made that was made." Thus, we can understand the emergent reality of the universe as narrated in this New Story of the universe. As cosmology it is even more appropriate for the abiding Logos presence and for the incarnation of Christ than any former mode of thinking about the universe.

Redemption: Our sense of entropy in the unfolding universe gives us a basis for appreciating sacrifice as a primary necessity in activating the more advanced modes of being. The first generation of stars by their self-immolation in supernova explosions shape the elements for the making of planet Earth and bringing forth life and consciousness. As with the great transformations of the non-living into the living, a sacrifice is made at one level of being to bring forth a more glorious reality. It is, then, consistent that the divine—which in a manner sacrificed its primordial unity in bringing forth the turbulent multitude of creatures—should enter into this sacrificial process even more profoundly in the redemptive deeds narrated in the Gospels.

Christian Community: The Christian community, too, can be seen as integral with the larger story of the universe, of Earth, of life, and of the human community. As Thomas states, the ultimate community is the whole universe together. This we can consider the primary intention of the divine in creation, redemption, and ultimate transformation. The Christian community in taking shape through history has a primary role in articulating this vision and achieving this purpose. Yet it must never be forgotten that the universe remains integral with itself throughout its extent in space and sequence in time. No subordinate segment of the universe community can replace this comprehensive community as the primary purpose or primary value of the universe. What is here said of the universe can also be said proportionately of Earth and all its components as a planetary community. Earth in its integral structure and functioning participates in the divine goodness more perfectly and represents it better than any single creature on Earth.

This integral universe, then, constitutes the sacred community par excellence and needs to be recognized as such in Christian thought. This realization is the basic reason for Christian concern for what is outside itself. The Christian community needs both to assimilate the rest of reality and to be assimilated by the rest of reality, by the human community, the Earth community, and the universe community, in accord with Saint Paul's expression of the divine being "all in all" (1 Corinthians 15:28).

5.

The Christian Future and the Fate of Earth

(1989)

The Christian future, in my view, will depend above all on the ability of Christians to assume their responsibility for the fate of Earth. The present disruption of all the basic life systems of Earth has come about within a culture that emerged from a biblical-Christian matrix. It did not arise out of the Buddhist world or the Hindu or Chinese or Japanese worlds or the Islamic world. It emerged from within our Western Christian-derived civilization. If these other civilizations were not ideal in their presence to the natural world, if they intruded extensively into the functioning of the planet, their intrusion, in its nature and in its order of magnitude, nowhere approaches the disturbance brought about by our modern Western disruption of the planetary process.

Although our Western industrial civilization was itself a deviation from Christian ideals, it came originally from within a Christian context. In its historical expression it could not have arisen out of any other tradition. We might conclude then that the Christian tradition is susceptible to being transformed in this direction. Until we accept the fact that our central beliefs carry with them a vulnerable aspect we will never overcome our present failure to deal with the increasing disruption of the planet.

If the planet fails then we fail, not only as Christians but even as humans.

That the planet is failing is evident from the meeting of biologists in September 1986 sponsored by the National Academy of Sciences and the Smithsonian Institution. At that meeting Norman Myers, the renowned Oxford biologist, stated that we are bringing about what may well be the greatest diminution in the variety and abundance of life on earth since the first flickerings of life almost four billion years ago. Harvard's E. O. Wilson indicated that we are probably losing some ten thousand species each year. Paul Ehrlich, the Stanford biologist, suggested that we may be bringing about through our industrial processes conditions similar to that of a nuclear winter. Later the following spring Peter Raven, director of the Missouri Botanical Garden, gave an address to the American Association for the Advancement of Science entitled "We're Killing Our World."

Whole volumes of such quotations from the best of our scientists could easily be assembled to indicate what is happening to the planet. Endless statistics could also be gathered to give evidence of the overwhelming destruction that is taking place. We need only read the annual report of the Worldwatch Institute, entitled the *State of the World*, or the biennial report of the World Resources Institute to fix the details in our minds. If we were totally realistic in training our youth for what is ahead of them, we would do so under the title: "Living Amid the Ruins"—amid the ruined infrastructures of the industrial world and amid the ruins of the natural world itself.

That Christians are ill-adapted to deal with these issues is clear from their general lack of concern for what is happening. Christians are somewhere off in the distance as, indeed, are most of the professions and institutions of our society. Probably in this country there is more understanding of the problem and there are more effective efforts at its solution outside the churches than within the churches.

I do not say that nothing is being done. There are a multitude of Christian environmental and ecological projects in this country and throughout the world. There is especially the

Au Sable Institute in Michigan, founded by Calvin DeWitt. There is the Web of Creation website on Christianity and ecology. There is the work being done by the Cathedral Church of Saint John the Divine in New York. Matthew Fox has for many years been guiding a program in creation-centered spirituality, a program that recalls with special emphasis those Christian traditions that experienced the divine in its cosmological manifestations. There are the efforts of those working toward an ecological balance through the ideals of an eleventh commandment. Genesis Farm is one of the foremost centers in the eastern section of the North American continent in guiding us toward an intimate human presence to the natural world. These are only a few of the Christian-sponsored projects that are functioning at the present time. My impression is that a new awakening is taking place within the Christian communities of this continent.

It is to foster these efforts and to make them more effective that I present these reflections. My first thought is that we not try to go it alone. The initial impulse of many Christians, it seems, is to go it alone, especially with those Christians most devoted to their religious heritage. There is the satisfying sense of bringing forth an ecology movement immediately derivative from their biblical-Christian heritage. This is important. It needs to be done.

Yet we have an even greater need to establish our identity with the more comprehensive ecology movements that are already far advanced in their understanding and efficacy, much further advanced than any of the movements that so far have emerged from a specifically Christian context. The ecology movement exists in its own right. It has inherently religious dimensions. It does not need biblical verification or consecration. Such piety, however valid, tends to alienate those whose rhetoric is different from the Christian rhetoric, and yet who feel the sacred dimension of Earth in the depths of its reality. Even when they make no reference to this sacred dimension and find it unsatisfactory to try to explain it, even then in doing the work itself they are fulfilling a sacred task. Their dedication has proven it. While some Christians were neglecting this task, others saw it and devoted an enormous amount of energy and talent to protecting the life systems of the planet. It is upon their foundations that we must proceed in the practical dimensions of our work.

That these other movements have a religious feeling about their work is often experienced as something of a threat to those Christians who are highly sensitive to what is sometimes considered naturalism, paganism, or even pantheism. When this attitude appears, the Christian presence is experienced by many ecologists as something alien or intrusive. In this situation the movement is divided. Its efficacy is diminished.

This difficulty could be mitigated if we were to recall that in earlier Christian ages the tradition considered that there were two revelatory sources, one the manifestation of the divine in the natural world and the other the manifestation of the divine in the biblical world. These needed to be interpreted in and through each other. In this context, to save Earth is an essential part of saving the pristine divine presence.

In the sixteenth century after the invention of printing, when written Bibles were more available, Christian emphasis became concentrated on the written text. The doctrine of the *two* books—the Book of Nature and the Book of the Bible—was diminished. The Book of Nature disappeared except for a few instances, such as in the work of the English naturalist John Ray, *The Wisdom of God Manifested in the Creation*, published in 1691. But even in instances like these we must feel that something is missing in the experience. Christians have been hesitant to enter profoundly into the inner reality of the created world in terms of affective intimacy. We do not hear the voices of the natural world. We tend to be autistic in relation to the non-human beings. We seldom appreciate natural reality in itself, a universe that unfolds from within its own powers, since divine reality does not make a world of automatons but a world of realities with real powers, even the powers of self-creativity. This is a dependent self-creativity that is further infolded in a divine presence that is, as Saint Augustine tells us, more intimate to us than we are to ourselves.

This hesitancy of Christians can be associated with the biblical experience of the divine as transcendent to all phenomenal existence, as personal, as creator of all that is, and as having established covenant relations with a chosen people. These are the teachings indicated in the opening chapters of the Book of Genesis and further emphasized in the first of the commandments given to Moses. In this context a sense of Earth and the pull toward an intimacy with Earth does not come easily for Christians; the more intense the Christian commitment, the more difficult such a sense, such a pull, is. The covenant itself, because of its juridic reference, has posed problems from the beginning, as we can see in the prophetic critique from Isaiah to Micah.

To this difficulty we must add the difficulties consequent on the emphasis given to redemption from a flawed world. Christianity is primarily a redemptive experience of salvation from the seductive forces within ourselves and those that surround us in the sensuous qualities of the natural world. This attitude creates an immense psychological barrier to our Christian intimacy with Earth. We are here, as it were, on trial, to live amid the things of this world but in thorough detachment from them. We long for our true home in some heavenly region. We long to pass over the river Jordan to that other world. That we truly belong to this world is difficult for us.

That the ultimate sacred community could be the universe itself is even more difficult, although such a proposal might be based on Saint Paul's Epistle to the Colossians where we are told that in Christ all things hold together and that we look forward to the time when God will be "all in all." As a further consideration we might note the great emphasis we place on the spiritual soul of the human, a soul created immediately by God, a soul that establishes the human in some manner above or outside the rest of creation. Every earthly reality is given over into our care. Thus we bear a responsibility for the community of creatures, since these glorify God in their own way. Yet there is an ambivalence in the relationship of humans with the natural world since the natural world exists also for the benefit of the human community.

Apart from the primary intention of the scriptures, the practice of Western Christians has been to consider that every earthly reality is subject to the free disposition of humans insofar as we are able to assert this dominion. We do not feel responsible precisely to the world about us since the natural world has no inherent rights; we are responsible only to the creator and to ourselves, not to abuse anything. But this leaves us in an alienated situation. We become an intrusion or an addendum to the natural world. Only in this detached situation could we have felt so free to intrude upon the forces of the natural world even when we had not the slightest idea of the long-range consequences of what we were doing.

Another observation concerning the difficulty of Christians in relating to Earth on any intimate basis has to do with the millennial vision of John in the Book of Revelation, a vision that promises a thousand years of intra-historical bliss toward the end of the world. The Great Dragon will be chained up. Christians will experience peace and justice and abundance and will no longer be subject to those basic life difficulties that we identify as the human condition. This promise of bliss in a transformed earthly context produces a radical dissatisfaction within the historical process, since every achieved form of fulfillment is insignificant in comparison with the promised preternatural fulfillment.

A final difficulty in our concern for the natural world is the prophetic message of care for the afflicted. Our finest achievement in the human order undoubtedly is our sense of the pathos of the human in both our individual and our social lives. Our ever more demanding social concerns, due to massive inequities and crippling poverty, may be one of the most difficult of the obstacles we need to overcome in establishing a more lasting human presence on Earth. To resolve this tension is one of the great challenges. It is now being called eco-justice, where social and ecological concerns are seen as deeply intertwined.

In overcoming these special ways in which Christians seem to be inherently distanced from Earth an ever-increasing emphasis has been given to the doctrine of stewardship.

According to this teaching based on the early chapters of Genesis, to us has been confided dominion over the earth and all its living creatures, "dominion" here interpreted as "care." While this teaching has its own attractive qualities and needs to be taught and practiced, we might ask if it fully resolves the tensions that we have here indicated. It does not seem to provide us with the feeling qualities needed to alter the destruction presently taking place throughout the planet. The doctrine of stewardship may be too extrinsic a mode of relating. It strengthens our sense of human dominance. It does not establish any intimate presence of ourselves to the world about us. Earth and ourselves: we remain separate and extrinsic to each other. Stewardship does not recognize that nature has a prior stewardship over us as surely as we have a stewardship over nature, however different the implications of these modes of stewardship. It does not enable us to overcome our autism at its deepest level. If we hear the voices of the natural world, these are not the fraternal voices heard by Saint Francis of Assisi but the voices of subservience.

Somehow there seems to be an affected quality in our efforts to establish an intimate presence to the natural world. In reality the story of the universe is our personal story, however we think of the universe. The reason for Christian aversion to the story of an emergent universe is that the story has generally been told simply as a random physical process when in reality it needs to be told as a psychic-spiritual as well as a physical-material process from the beginning. In my view, the first step in achieving any adequate human or Christian activity in saving the planet from further irreversible dissolution is to recognize that the universe story, the Earth story, the life story, and the human story—all are a single story. Even though the story can be told in a diversity of ways, its continuity is indisputable.

We need only reflect on the fact that any diminishment of the splendor of Earth is a diminishment of the human in its most sublime functioning. As Saint Paul tells us in the first chapter of the Epistle to the Romans, the invisible divine reality is known from the visible things that are made. We have a glorious sense of the divine because we live in such a magnificent world. If we lived in a less resplendent world, our sense of the divine would itself be diminished. As we lose our experience of the songbirds, our experience of the butterflies, the flowers in the fields, the trees and woodlands, the streams that pour over the land and the fish that swim in their waters; as we lose our experience of these things our imagination suffers in proportion, as do our feelings and even our intelligence. If we lived on the moon, our sense of the divine would reflect the lunar landscape, our imagination would be as empty as the moon, our sensitivities as dull, our intelligence as limited.

Not even the scriptures could replace what we would have lost, nor could the incarnation itself or redemption or the prophetic teachings concerning care for the afflicted. Our rootedness in Earth is itself a condition for any of these things taking place. None of it has any adequate meaning apart from the basic structures and functioning of all those glorious and nourishing forms that surround us, all those stupendous natural experiences that we have in the dawn and sunset, in the seasonal sequence, in the rainstorms of summer and the blizzards of winter.

We are so integral with the world about us that we might consider the universe itself as the larger dimension of our own being. We, in turn, enable the universe to reflect on and celebrate itself and its numinous origin in a special mode of conscious self-awareness. The universe has what might be considered a human dimension from the beginning. Even in modern cosmology, when considering the place and role of the human in the universe some of our best physicists have evolved what is known as the anthropic principle. According to this principle the universe must, from its beginning, have had tendencies that would eventually lead to the emergence of the human. The integrity of the human with the natural world might be accepted regardless of how a person views the question of the beginning, whether in accord with historical developmental processes or through such sequences as are indicated in the Mediterranean stories of creation.

In 1988 one of the most profound statements of concern for Earth from within a Christian institutional context was the pastoral letter of the Philippine bishops entitled "What Is Happening to Our Beautiful Land." After describing the devastation being inflicted on the land, this pastoral letter tells the Philippine people: "At this point in the history of our country it is crucial that people motivated by religious faith develop a deep appreciation for the fragility of our islands' life systems and take steps to defend the Earth. It is a matter of life and death." [See their statement in 2008 also.]

Seen in its larger context, the urgency of the work to be done here in our own country and throughout the world requires a coherent effort of the human community across all the boundaries of nationality, ethnic derivation, cultural formation, and religious commitment. A comprehensive expression of this task and the means of its fulfillment are contained in the Charter for Nature, passed by the United Nations General Assembly in 1982. There, a clear recognition was given to the dependence of all human affairs on the larger context of the natural world:

> Mankind is a part of nature and life depends on the uninterrupted functioning of natural systems which ensure the supply of energy and nutrients. Civilization is rooted in nature, which has shaped human culture and influenced all artistic and scientific achievement, and living in harmony with nature gives man the best opportunities for the development of his creativity, and for rest and recreation ... Every form of life is unique, warranting respect regardless of its worth to man, and, to accord other organisms such recognition, man must be guided by a moral code of action.

This charter also calls for every person in the world to assist in fulfilling the directives of this charter.

It was estimated a few years ago that there are in this country some twelve thousand organizations, movements, and publications devoted to defending the environment from further destruction. My own feeling is that this number has considerably increased in more recent years. Indeed, it might be said that the most significant division among humans at the present time is based neither on nationality, ethnic origin, social class, or even religious commitment, but is rather the division between those dedicated to exploitation of Earth in a deleterious manner and those dedicated to preservation of Earth in all its natural splendor. This difference might be phrased somewhat differently by using the terms "anthropocentric" and "biocentric" as the basic referent as regards reality and value.

Religion undoubtedly would already be deeper into such issues except for the pressing social issues and the pathos of the human in contemporary urban society. But now the loss of topsoil, the poisoning of the air and water, the threat of the weakening ozone layer, the increasing greenhouse effect on our climate, the extinction of so many species of plants and animals, the elimination of forest lands, the spoiling of our beaches—all these things are drawing us ever more deeply into issues that we have never dealt with previously, issues that must be resolved in some effective manner if our Christian social programs are to have any lasting efficacy.

Morally we have a well-developed response to suicide, homicide, and genocide. But now we find ourselves confronted with biocide, the killing of the life systems themselves, and geocide, the killing of the planet Earth in its basic structures and functioning. These are deeds of much greater evil than anything that we have known until the present, but deeds for which we have no ethical or moral principles of judgment.

The changes we are bringing about in their nature and in their order of magnitude are the most significant changes that have taken place since the modern human came into existence some sixty thousand years ago. These are not simply cultural changes, such as the change from the classical Roman period to the medieval Christian period or the change from the medieval to the Renaissance period. The changes presently taking place are changes in the chemical constitution of the planet, in its geological structure and in its biological systems. We are eliminating life forms that took hundreds of millions, even billions of years to bring into existence. The tropical rainforests took sixty million years to come into existence. They are possibly the most beautiful life expression on Earth or even in the universe. Yet we are wiping out these rainforests at the rate of fifty acres every minute of every day. A simple doctrine of stewardship does not seem adequate in dealing with such massive issues. More profound developments in our sense of relatedness to the natural world are demanded.

In conclusion, I propose that new religious sensitivities need to be developed. In former times if such a situation had existed, a new religion might have arisen. But the time is over, apparently, when a religion like any of the classical religions could come into being. What is needed now is not exactly a new religion but new religious sensitivities in relation to planet Earth that would arise in all our religious traditions. The model for these new sensitivities might well be the sensitivities that can be observed in the earlier, more primordial religions, sensitivities that can still be found among some of the tribal peoples of the world.

While these traditions are sometimes considered primitive nature religions possessing none of the grandeur or authenticity of the Christian religion with its transcendent monotheistic personal deity/creator of a world clearly distinct from himself, still they contain insights into the basic relations of humans to the natural world that we are desperately in need of just now and that we cannot articulate within the context of our own resources. The most needed of these insights is the realization that humans form a single community with all the other living beings that exist upon the earth.

In accord with the teachings of Saint Paul and Saint John we might perceive that there is a Christ dimension to this more extensive community of Earth and that what we do to this community we do in some manner to Christ himself. It is difficult to believe that God created such a beautiful world if it were not also the divine intention to redeem, sanctify, and bestow eternal blessing upon it throughout eternity.

6.

The Role of the Church in the Twenty-first Century

(1995)

Some years ago Peter Raven, the director of the Missouri Botanical Garden, gave an address to the American Association for the Advancement of Science entitled "We're Killing Our World." We have been told similar things many times by some of the most distinguished biologists of our times, including E. O. Wilson and Norman Myers, biologists with comprehensive knowledge of the life systems of the planet.

More people are becoming aware of this at the present time. Yet the thought is so awesome, the demands for remedy so absolute that we have entered a state of denial. My own generation, the generation that was born early in the twentieth century, I consider to be somehow autistic. They have had no intimate rapport with the natural world. They could not have devastated this continent so terribly had they possessed the slightest sensitivity for the wonder of the world they were bringing down in ruins.

When we talk of the role of the Church in the twenty-first century, we must begin with the obvious reality of an industrial civilization ruining the natural world on which we depend for both our physical and our spiritual sustenance. The basic question is no longer human-divine relations; nor, in my view, is it inter-human relations. The basic issue is human-Earth relations.

The future of the other two relations depends upon this third relation, our human capacity to recognize our place in the structure of the universe and to fulfill our role within this setting.

What is happening now is something far beyond any historical change or cultural modification that humans have known in the past, changes such as occurred when we moved from the classical to the medieval period or from the medieval world to the Enlightenment period. Both in its modality and in its order of magnitude what is happening now is something vastly different.

We are changing the chemistry, the geosystems, and the biosystems of the planet on a scale such as has not occurred for the past sixty-five million years, since the beginning of what is known as the Cenozoic era. Already we have terminated this era in its basic creativity. The only viable future for the planet or for ourselves is to recognize the devastation that we have caused and enter a new era that we might think of as an Ecozoic era, a period when humans would be present to the planet in a mutually enhancing manner.

There is one simple cause for the devastating situation in which we find ourselves. We have replaced the universe as the primary referent concerning reality and value in the phenomenal world with the human as the supreme referent of reality and value. We have rejected the divinely established order of the universe and are attempting to establish a contrived human order in its place, under the assumption that we know better than nature how the universe and planet Earth should function. No pathology ever invented could be so perverse and so devastating to the delicate balance of life and existence on this planet.

Along with this distortion of the order of the universe we have broken the unity of the universe and especially of the life systems on Earth. We have established a discontinuity between the non-human and the human components of the universe and have given all rights to the human. We have considered the universe as composed of objects to be exploited rather than as subjects to be communed with.

All rights have been given to the human. The non-human has no rights. All basic realities and values are identified with the human. The non-human attains its reality and value only through its use by the human. As economist Peter Drucker insists: "Before it is recognized for some use, every plant is just another weed." This deep cultural pathology in Western society has been communicated throughout the planet, letting loose the devastating assault on the planet that we witness everywhere.

If this situation has come about through our assumption of a radical discontinuity between the non-human and the human orders of reality and the denial of inherent rights to the non-human, then the solution for this pathology is our recognition that the primary value rests, not with the human, but with the larger community within which the human comes into being. This larger community has given to the human everything that the human possesses, both physically and spiritually. The special gifts bestowed upon the human are given, not primarily for the human, but for the perfection of the entire universe.

That perfection requires the human to fulfill a certain role of enabling the universe to reflect on itself and that numinous origin from which the universe and every being in the universe comes into being, is sustained in being, is moved to action, and is brought to its fulfillment. Humans exist for the integral community of existence. This integral community that we refer to as the universe is in the phenomenal order the efficient cause and the final cause as well as the formal and material cause of all that is human. The realization of this makes the assertion of the human as its own major concern inadequate. The ultimate concern of the human must be the integrity of the universe upon which the human depends in such an absolute manner.

If we read our present historical situation in this context, then what is needed is an adjustment of the human to the universe. The basic issue before us is not human-divine relations or inter-human relations but human-Earth relations. This adjustment on our part to the realities of Earth requires that we move into a new geo-biological period in the story of our planet. We need to move from the terminal Cenozoic to what might be designated as the emergent Ecozoic era in Earth history, the period when humans would be present to the planet in a mutually enhancing manner. We need to establish ourselves in a single integral community including all component members of planet Earth. This conclusion follows the reasoning of Saint Paul in the twelfth chapter of his First Epistle to the Corinthians.

This broader integral community must be the primary concern in all human affairs before that of any particular member of the community. Even our sense of the divine comes to us mainly, not through any scriptural communication, but through the universe around us. The natural world is the fundamental locus for the meeting of the divine and the human. We need to look up at the stars at night and recover the primordial wonder that awakened in our souls when first we saw the stars ablaze in the heavens against the dark mystery of the night. We need to hear the song of the mockingbird thrown out to the universe from the topmost branch of the highest tree in the meadow. We need to experience the sweetness of the honeysuckle pervading the lowlands in the late summer evening.

In all these experiences communion takes place between ourselves and that numinous reality whence the universe came into being and by which it is sustained in its immense journey. For too long we have had these experiences only in our imagination, and with our souls closed to the mystery they are revealing. If we were truly sensitive, we would realize that in these moments the universe is communicating to us the most basic understanding that we really need.

The difficulty is that we are no longer responsive to such experiences. Scientists have taught us that the universe is a random process without inherent meaning. This attitude has locked us into ourselves so tightly that we cannot get out of our human selves into the larger universe, nor can this larger universe enter our own inner world or establish any intimate rapport with us. This pathology is the ruin of both the planet and ourselves. It is a primary cause for the devastation of Earth.

That we are awakening from this trance is due to the forebodings we are experiencing as the most beautiful of Earth's living forms are disappearing, as our children grow restless and soul-hungry for a world of meaning beyond anything available from their concrete world, with all its noise and trash and wires and wheels and machinery and electronic gadgetry.

Yet no remedy can be effective until we accept our place and our role in the universe, until we recognize that other creatures of Earth have rights to their lives, to their habitats, to their own fulfillment and the opportunity to carry out their role in the larger functioning of the universe. To understand and accept our own limited role and our own limited rights in the world about us is a first step that we must take.

Earlier in the course of human affairs this communion with the universe and with the spirit-presence in the universe took place spontaneously. The grandeur of the mountains was a spiritual mode of being. Sunrise and sunset were sacred moments. Animals were spirit presences. The human mind was awakened to beauty. An enduring intimacy was established between the human and non-human worlds. All human affairs were understood within the functioning of this larger community of existences. Religious ritual, prayer, poetry, and music were born of this source. The primary human obligation was integration into this larger structure and functioning of the universe as a sacred mode of being. Acceptance of this fact was the foundation of the profound wisdom possessed by indigenous peoples the world over.

Intimacy with the natural world is evident, for example, among Dineh/Navaho peoples in the southwest of the North American continent. In describing one of their central religious ceremonials, Blessingway, the Navaho elder, River Junction Curly, said: "With everything having life, with everything having the power of speech, with everything having the power to breathe, with everything having the power to teach and guide, with that in blessing we will live..."[1] And again, a Koyukon elder from these northern Alaskan indigenous peoples remarked: "Each animal knows way more than you do. We always heard that from the old people when they told us never to bother anything unless we real-

---

[1]James K. McNeley, *Holy Wind in Navajo Philosophy* (Tucson: University of Arizona Press, 1981), 28.

ly needed it."[2] While it is not necessary to equate these statements, a shared ethic is obvious, and that message is that life in the immediate cosmos, namely, the local bioregion, is alive in numinous ways. Indeed, for these elders the community of life both blesses and calls forth responsibility from humans.

So, too, in the greater civilizations the primary structures of human life were modeled on respecting the structures and functioning of the universe. All were governed by a cosmic reference, such as the *rita* of Hindu India, or the *tao* of China, or the *kosmos* of Greece. The rhythms of human life were coordinated with the rhythms of the natural world, primarily with the cycle of the seasons. The great rituals of Mesopotamia were re-enactments of the creation event. According to the sixth century BCE Chinese classic, the *Tao Te Ching*, "Humans model themselves on Earth, Earth models itself on Heaven, Heaven models itself on Tao, Tao models itself on its own spontaneities" (chapter 25). In the Jewish world, the sense of the law of the Lord had a cosmic dimension. In the most ancient Christian liturgy, the Easter Vigil, the creation story constituted the context of the entire redemption experience.

This primacy of the universe was transmitted in Christianity through the *Hexaemeron* literature, that is, the commentaries written to explore the six days of creation described in the first chapter of Genesis. Thomas Aquinas received not only this commentarial tradition but also the reintroduction of Aristotelian thought with its emphasis on form in matter. Thus, Thomas affirms the vision in Genesis of God looking over the entire community of life after the sixth day of creation and saying it is "very good."

In all these instances there is a realization that the universe is the encompassing context of everything that happens in the universe. The universe itself is recognized as the only self-referent mode of being in the phenomenal world. Every other mode of being is universe-referent. The universe is primary law-giver,

---

[2]Richard Nelson, *Make Prayers to the Raven: A Koyukon View of the Northern Forest* (Chicago: The University of Chicago Press, 1983), 225.

primary educator, primary revelation of the divine, primary economic resource, primary engineer, and primary artist—indeed the primary agent in every aspect of existence within the phenomenal order. This order includes, of course, planet Earth and all phases of life that have emerged on Earth.

We are not living in any of the ancient civilizations, nor have we participated in their effort to build a significant human mode of life within the cosmological order. Neither are we people of the Western medieval period with their sense of the order of the universe as the context in which human affairs have their proper significance. Yet it is necessary to speak of these earlier traditions in order to identify the continuing need that humans have for placing human affairs within the integral structure and functioning of the universe. All these earlier traditions had two sacred obligations. One was to understand and revere the all-pervasive numinous presence throughout the universe. The other was to realize the essential continuity of all things within the universe and to abhor any isolation from the larger community of existence.

Both these commitments of earlier human societies have been abandoned, and we have, as a result, become lost and the world about us desolate. A break of immense significance has occurred. We have moved from an earlier mythic mode of understanding the universe to an understanding based on empirical processes and expressed in numerical equations. We have invented instruments that enable us to see and hear across great distances in space and time and by these means we have entered more intimately into the physical order of the universe both in its microphase and its macrophase structures. We have made prodigious discoveries altering forever our relation both to Earth and to the universe. By these means, also, we have gained control over Earth to an extent and depth never dreamed of in former times.

The greatest single discovery that we have made is that the universe is an emergent process that has come into its present mode of being through a long sequence of irreversible transformations over a period of some fourteen billion years. We have discovered the universe story. This story is the story of each being in the universe and all modes of being in the universe. It is the story of ourselves individually, but also of ourselves as a single comprehensive yet intimate community. There is ultimately only one story, the Great Story.

The intimate, genetic relationship of the human with the non-human world is now known to us with new clarity and comprehension. We understand especially that each being in the universe is what it is only because every other being is what it is. Nothing is itself without everything else. There is no discontinuity. There is distinction. There is no separation. Otherwise the universe ceases to exist. Throughout its vast extent in space and its sequence of transformations in time the universe remains a single multiform reality. If we could make the continuity of the non-human with the human the effective operating principle in all our human institutions, professions, programs, and activities, we would soon bring healing to this damaged planet.

At this moment of transition the twenty-first-century Church, which has lost a sense of its basic purposes in these past centuries, could restore its efficacy and extend its influence over human affairs. The Church could be a powerful force in bringing about the healing of a distraught Earth. The Church could provide an integrating reinterpretation of our New Story of the universe. In this manner it could renew religion in its primary expression as celebration, as ecstatic delight in existence. This, I propose, is the Great Work to which Christianity is called in these times.

In relation to its traditional teaching, a new understanding will need to emerge in almost every aspect of belief, discipline, and worship. In most instances this change will lead back to fundamental emphases too long neglected. It should first be observed that our new understanding of the universe through empirical processes and through our instruments of observation has given us an immediacy—even a new intimacy—with the universe in its sacred dimension. We now see the universe as a primordial flaring forth of immense energies that soon moved from an undifferentiated radiation into differentiated forms of matter with three basic tendencies: toward differentiation, toward inner spontaneity, and toward a comprehensive bonding with each other. The first atomic forms to take shape were hydrogen and helium, the simplest elements. These constituted the first stars that, in galactic shape, provided the primary pattern of the universe from its earliest times until the present.

This primordial flaring forth must have been an event gorgeous beyond description. We can only marvel as we reflect that we ourselves and the world that we observe about us are the explicated forms of what was implicated in that first flaring forth of the created world. The entire story of these fourteen billion years between this beginning and our present situation is the supreme epic of our times, an epic with a grand sweep beyond anything even imagined in previous times.

This primary emergence of the universe and the long sequence of its transformations until the present, as we now know them, constitute a new revelatory experience of that numinous source whence all things came to be in the beginning, its power and its grandeur, but especially of that comprehensive embrace whereby the divine holds all things in unity. To speak at this point of revelation is quite appropriate, although the term, meaning an unveiling, is here used in a sense qualitatively different from that expressed in biblical exegesis.

We might now reflect on both the differentiating and the containing forces of the universe from the beginning and their relationship. These two forces must exist in a certain balance, for if differentiating forces overcome containing forces, the universe would explode at the first moment and nothing else would happen. If, on the other hand, containing forces overcome differentiating forces, then the universe collapses and again nothing happens. Should the two forces come to perfect equilibrium, the universe fixates, and, yet again, nothing happens. The only possibility for an emergent universe requires that a creative disequilibrium be established so that an unfolding process can take place throughout the future. This was in fact what happened in the case of our universe. In the course of evolution the creative disequilibrium found expression in the curvature of space that holds all created reality in a single bonding, a single embrace. All intimate relationships begin here. Here is the origin of gravitational attraction whereby every physical being attracts and is attracted to every other physical being.

I refer to this bonding force of the universe as the Great Compassionate Curve, for this curve, as every other mode of being of the universe, has a psychic-spiritual aspect as well as a physical-material aspect. The curvature of space expresses in our times that perception of the universe that in late Paleolithic times was expressed by the Mother Goddess as the nurturing principle that brought the universe into being and enabled Earth to bring forth such abundance of life. This fertility figure exists in our times in the image of a simple curve whereby the universe is held together in all those intimacies whence life emerges in such abundance.

Another remarkable aspect of this primordial moment of the universe is that the entire future of creation emerged out of the three governing tendencies—differentiation, subjectivity, and relationality—that appear at this time. While all three tendencies appear immediately, we might speak first of the tendency toward differentiation. Each being in the universe is unique. Each being gives to the universe something that no other being can provide. Thus the inherent value of each reality. Each leaf is different, as is each snowflake, each flower, each dawn, each sunset. At the human level, each individual seems to be almost a different species.

The second tendency, namely subjectivity, is the emergence of inner articulation. Each being has its own interior structure, its intelligibility, its inner spontaneity, its voice. Each individual being has its inner principle of action, its capacity to relate to every other being in the universe. Each being shares to some extent, however attenuated, in the qualities of consciousness and personality. Nothing is simply an object to be used. Each being has its dignity calling forth esteem. Each being carries the presence of the divine.

The third tendency of the universe is that of relationality or communion. This is the tendency toward the bonding of each being with every other being in the universe. Nothing can ever be separated from anything else. We are most fully ourselves in communion with other humans and with all the living beings of Earth.

It should be immediately evident that this threefold tendency provides a remarkable model for understanding the Christian doctrine of the Trinity. We have the family model presented in the Bible as Father, Son, and Holy Spirit. We have the psychological model presented by Augustine as intellect reflecting on itself. We have the social model sometimes used in modern times as the self, the other, the community. Yet none of these examples has the special qualities of the cosmological model, which present the Father as the principle of Differentiation; the Son as the icon, the Word, the principle of inner articulation; the Holy Spirit as the bonding force holding all things together in a creative, compassionate embrace. In this context the Church should have little difficulty understanding the universe as the primary sacred community.

In a similar manner the incarnation can be seen in a cosmological context. Here we find a profound way in which our new understanding of the universe provides an unsuspected depth of insight into incarnation in accord with the teachings of Saint Paul in the Epistle to the Colossians and Saint John in the preface to his Gospel. Neither of these scriptural writers saw the Christ figure simply as an individual. The Christ event was of cosmic dimensions, requiring a cosmic as well as an individual mode of being. This is why Saint Paul tells us that in Christ all things hold together. In like manner, John begins his Gospel not immediately with the human birth of Christ but with the Eternal Christ as the creative Logos of the Universe.

If we understand the Christ reality to have a cosmic aspect, so we must consider that the universe has a Christ aspect. This is the Christian mode of understanding the universe itself in terms of the human form. This sense of cosmic person is extensively used in the traditions of various civilizations. It occurs with special clarity in the Buddhist world with the cosmic or ontological dimension of the Buddha figure. It occurs in the *Mahapurusha* (Great Person) concept of Hinduism, as well as in the concept of the one body of the universe in the Chinese tradition. Amazingly, this sense of cosmic person occurs in the most advanced modes of interpretation of the universe as expressed by modern physics.

There it is spoken of as the "cosmological anthropic principle." In this context the relationship between the human mind that knows the universe is considered in its relation to the universe whence the human mind comes into being.

Here we meet with statements such as that of Freeman Dyson who says that the more he considers the structure of the universe and its functioning the more evident it seems that the universe must have known that we were coming. While avoiding the type of predeterminism fostered in some religious traditions, this conclusion does reflect a certain tendency in the universe toward reflection on itself in the human mind. While such ideas do not immediately refer to the cosmic dimension of the Christ reality, they do indicate that we cannot consider the universe without in some manner relating it to the cosmic dimension of the human. In the Christ appearance a more explicit manifestation of the numinous dimension of the universe is found.

We know the universe through its varied modes of expression. Since the universe has expressed itself in beauty and life and consciousness, then the universe must be a beauty-producing, life-producing, and consciousness-producing process. With such capacities we can only wonder at the numinous reality, the divine reality, whence came the primordial blazing-forth and the long sequence of transformations that have occurred until the present.

The role of the Church in the twenty-first century, then, is to speak more directly concerning the universe itself as the primordial revelation of the divine. It should be clear that verbal revelation cannot be a primary revelation, since any communication that takes place through language takes on the distortions of the language, the particular social forms of the times, and the complex patterns of historical events occurring during that period. In contrast, the revelation of the natural world directly and immediately awakens a sense of awe and mystery along with a sense of creatureliness. It arouses, as well, a tendency to worship.

Until recently, Christians have been concerned with the soul, the inner life, spiritual disciplines, sacramental participation, works of mercy, care for the impoverished—activities all leading to salvation in a trans-earthly realm. However unfortunate the loss of the natural world in all its grandeur, this, they contended, was not essential to the spiritual salvation of the soul. It was merely the realm of the secular naturalist or the mechanistic scientist. In this manner the discontinuity between the human and the non-human was profoundly affirmed within Christianity.

To counter this tendency we need a more adequate understanding of the universe, of how it came into being, of its governing tendencies, and of the sequence of transformations whereby it has taken on its present forms of expression. We need to know how the solar system and Earth came into being, how life developed on the planet, and, finally, how we ourselves appeared and what our human role has been within this amazing process. All these things need to be understood as aspects of a spiritual as well as a physical process. Only such comprehensive and deep understanding can restore the integral functioning of planet Earth upon which human well-being depends. This is the fundamental task of the Church in the twenty-first century.

7.

Christianity and Ecology

(1997)

The survival of planet Earth in its integral reality is, it seems to me, the basic issue that confronts us here. In some basic sense the human project and the Earth project are a single project. There is no way in which the human project can succeed if the Earth project fails. That this conclusion is not understood by the guiding forces of the human community is the challenge that is before us.

That the Earth project is failing just now seems obvious, both to ordinary observation and to more scientific understanding. We are at the terminal phase of the Cenozoic period. Referring to species extinction taking place in our time, biologists tell us that nothing of this order of magnitude has happened in the geo-biological sequence of life on Earth since the beginning of the Cenozoic age some 65 million years ago. Warning reports have also come from the world's foremost nature organizations concerning the status of plant species. One in every eight of these throughout the world is imperiled. In

A version of this essay was published under the title "Christianity's Role in the Earth Project" in *Christianity and Ecology*, ed. Dieter Hessel and Rosemary Ruether (Cambridge, MA: Harvard Center for the Study of World Religions, 2000).

America almost one in three is endangered. Obviously, then, the larger destiny, not only of the human but even of more extensive realms of existence is being determined.

We need to grow accustomed to thinking in terms of the Cenozoic period, for this is Earth's great lyric period in the sequence of planetary life development. It is also the period when humans emerged, profoundly conditioned by life forms and other environmental factors that developed during this time. Our inner world of genetic coding, shaped by the same forces that created the world about us, is integral with this outer world. Even our soul life is developed only in contact with these surrounding experiences. So integral is our inner world with the outer world that if this outer world is damaged, then the inner life of our souls is diminished proportionately. We are genetically coded for existence in that late Cenozoic world in which experiences narrated in the Book of Genesis and other biblical texts took place. The conclusion? When we so ruthlessly extinguish the life forms of our period, we threaten, along with these planetary beings, the inner life of the human.

To preserve this sacred world of our origins from destruction, our great need is for renewal of the entire Western religious-spiritual tradition in relation to the integral functioning of the biosystems of planet Earth. We need to move from a spirituality of alienation from the natural world to a spirituality of intimacy with it, from a spirituality of the divine as revealed in verbal revelation to a spirituality of the divine as revealed in the visible world about us, from a spirituality concerned with justice simply to humans to a justice that includes the larger Earth community. The destiny of Christianity will be determined to a large extent by its capacity to fulfill these three commitments. My purpose here is to identify the present status of Christians in regard to these issues and to propose orientations that might assist in dealing with the situation.

As Christians we lost our intimacy with the natural world, it seems to me, in three phases. The first phase occurred during the meeting of early Christian spirituality with Greek humanism to form the basis of a strong anthropocentrism. This anthropocentrism would in the course of the centuries so exalt the human as to lose the sense of the human as only one component in the larger community of existence. In addition to the influence of Greek humanism, biblical revelation itself has overwhelmed the revelation of the natural world and concern for the pathos of the human has left little energy to care about the nonhuman world. In this sense, we need a new integration of human-Earth concerns in eco-justice. In addition, the intensity of human spiritual commitment to the divine served to weaken concern for the experience of wonder and awe generated by the natural world. That world would be experienced as a distraction from higher things.

The second phase in the alienation of humans from the natural world came about when the Black Death ravaged Europe from 1347 until 1349. In this devastating period of Western civilization, a third of Europe died. The plague was especially severe in Florence where, in the summer of 1348, fewer than 45,000 persons of the 90,000 living there at the beginning of the year survived. In Siena of a population of 40,000 only 15,000 persons survived.

The difficulty was that people had no explanation for what was happening. They could only conclude that the world had become wicked and that God was punishing the world. The great need was for repentance, withdrawal from the world, and an increasing quest for redemption. A spirituality developed that involved disengagement and even abnegation from worldly concerns. This spirituality of detachment found expression in a new devotional intensity directed toward the Savior personality. In Florence, Dominican preacher Jacopo Passavanti taught a severe doctrine of penance and discipline in relation to the natural world.

This was also the period of the *Ars Moriendi*, a kind of spiritual guide to assist and console the dying. Some forty manuscripts of this treatise still survive. The late fourteenth and early fifteenth centuries saw the "Dance of Death" emerge as a common theme for art. In addition, austere morality plays were composed and presented extensively throughout England. The theme of these plays was the battle between good and evil for the human soul; the basic principle was that nothing is worthwhile, and that we cannot take anything with us when we die except the merit of our virtues. Everything else is a kind of vanity. The supreme morality play in fifteenth-century England was *Everyman*, a play significant in the origins of English drama. Art, too, was notably affected. For the first time there was emphasis on Last Judgment scenes depicting Christ with upraised right hand condemning the wicked to hell. Hell at this period became an artistic subject with lurid descriptions of decay and death and eternal misery.

The consequences of the Black Death can thus be seen in the development of a spirituality of detachment from worldly concerns. This spirituality found expression in the work of Thomas à Kempis, whose fifteenth-century book, *The Imitation of Christ*, was to become a classic of Christian spirituality over the next five centuries until the mid-1900s. The spirituality of detachment led to the Puritanism of the Protestant world and the Jansenism of Catholics.

The post-Black Death period, then, saw the Western soul positioned in radical alienation from the natural world. Instead of delight and a pervasive experience of the divine in the world's beauty, wonder, and awesomeness, there developed a psychic-spiritual withdrawal from too intimate relations with one's surroundings. This attitude has continued until the present day, with a certain relief at the end of the eighteenth and the first half of the nineteenth centuries when the Romantics once again exalted the natural world as the abode of some higher, numinous spirit. The inevitable reaction led to realism and eventually to the existentialism of the mid-twentieth century.

Sixteenth- and seventeenth-century alienation from the natural world left the scene open for scientific perception of the world as mechanistic in structure and subject to whatever technological controls humans could invent to make nature increasingly useful to them. In some sense it could be said that science and technology took over and exploited the planet because religious persons had abandoned it.

A third moment in our loss of intimacy with the natural world occurred at the end of the nineteenth century when we abandoned our role in an ever-renewing, organic, agricultural economy in favor of an industrial, extractive economy. This development decisively moved the scientific and technological might of the modern world into a merciless program of disruption of the organic functioning of the planet. At this time the planet lost its wonder and majesty, its grace and beauty, its life-giving qualities. It became an object of *use*, of exploitation. No remnant of its subjective qualities was permitted to remain. Engineers took over—the mechanical, the electronic, the chemical, and now the genetic engineers. These persons flaunted their technical competence and asserted their intellectual arrogance in assuming that the planet would be better off under their control than under the primordial control of those forces that first brought Earth into being and guided its development down through the ages.

As a result of these three phases of alienation from the planet—Greek and biblical anthropocentrism, the Black Death, and the movement to a technological and industrial economy—the profound spiritual aspect of Earth has been almost completely negated. The radiant presence of the divine is no longer recognized. Any awareness of spiritual communication made by the natural world to the human has disappeared. The universe has become a collection of objects, not a communion of subjects. Even the most sublime realities have become susceptible to economic exploitation. Planet Earth has become a commodity to be bought and sold, not a place where humans find both physical and spiritual nourishment. The basic problem before us now is how to recover a sense of the universe as manifestation of some numinous mode of being.

We cannot save ourselves without saving the world in which we live. There are not two worlds, the world of the human and a world of the other modes of being. There is a single world. We will live or die as this world lives or dies. We can say this both physically and spiritually. We will be spiritually nourished by this world or we will be starved for spiritual nourishment. No other revelatory experience can do for the human what the experience of the natural world does.

When we try to understand the universe and our place in it, we find ourselves doubly estranged, not only from the universe of Genesis but also from the universe as we now know it through empirical observation and scientific insight. Spiritually we may feel ourselves at home with the Genesis story. Yet, as regards understanding the evolution of the universe, its emergence into being, the sequence of transformations through which it has passed, and the manner of its functioning, Genesis cannot enlighten us. At the same time, the scientific story of the universe, such as we now know it, is communicated to us as merely material in its substance, mechanistic in its functioning, and random in its development.

Once we get beyond the explanation given by scientists, however, and look at data derived from their intense observation of the universe over these past few centuries, we begin to see a story of immense significance, a story that reveals the deepest of mysteries if only we know how to understand the story. We now perceive the universe as having come into being from an original flaring forth of primordial energy, then passing through a sequence of irreversible transformation episodes that have brought into being the visible world about us.

This phenomenal universe that we observe in such detailed scientific fashion cannot be explained simply in itself. In every way it is dependent on a numinous, trans-phenomenal, divine creative power. When we explain the universe as we now know the universe in its originating moments and its long sequence of transformations, we are explaining the manner in which the Creator has brought the universe into being. Yet we must recognize that the universe is not a puppet creation manipulated by some transcendent power. Such a creation would not satisfy the purposes of creation such as we know these to be. The created being would not have the independence needed to enter into an intimate divine-human relationship. Such a being could not freely offer divine praise. For these reasons, the divine creates a phenomenal world with the power to develop greater complexity through emergent processes. The wonderful thing about the universe is that it constitutes an absolute unity in which each component is universe-referent and all the components are inter-referent among themselves.

This universe, which we must now understand as our sacred universe, is the same universe as that presented in the Book of Genesis. It is a universe, however, that is experienced through immediate empirical observation rather than simply through the inspired words of a narrator writing in a distant region and an ancient time in a strange language. Through this observational process we have also come to know the universe as an emergent process over an immense period of time. Once this emergent reality is seen within a religious and spiritual context, once it becomes clear that the universe itself has a spiritual dimension from the beginning, then we have the basis for a new cosmology in which we can find new depths of meaning in biblical truths.

Here it is important to appreciate the role of the universe in the Christian context. So estranged from the universe have we been over these past centuries that we feel Christian spiritual tradition is independent of any need to be concerned about the universe. So long as we maintain an intense belief in Jesus, so long as we develop our interior intimacy with the divine, so long as we follow the Christian life discipline, so long as we carry out the spiritual and corporal works of mercy toward others, so long as we focus our lives on the Gospel, then any concern about the universe or planet Earth has no great urgency, no overwhelming relevance to the spiritual process. If others take over the planet to exploit its resources or even to cause severe disruption in the life systems of the planet, this devastation is not of great personal or Christian concern.

So, too, is the case with ministry. The Bible is perceived as the basic context of belief and the way to salvation. To make events occurring in the natural world a matter of discussion can distract. The proper role of Church homilies is primarily to concentrate on Gospel narratives telling of the activities and teachings of Jesus.

In considering the relevance of the larger world to the Christian tradition we might note that apparently no divine communication could have been given to humans until the cosmological setting had been established. Once the creation narrative had been presented in the Book of Genesis, scriptural events could be understood as embracing the universe and specifically the Earth community. In the Psalms we find that divine praise is inspired by the entire universe and constitutes the great cosmic liturgy. Psalm 104 says of Earth and its creatures, "When you send forth your spirit, they are created; and you renew the face of the ground" (Psalm 104:30). Saint John tells us that "God so loved the world [*kosmos*] that he gave his only Son" (John 3:16) that the world might be redeemed by him. Saint Paul writes that "in Christ all things hold together" (Colossians 1:17).

Thus the basic Christian understanding of the universe is one in which the human community and the natural world are seen as a unified, single community with an overarching purpose: the exaltation and joy of existence, praise of the divine, and participation in the great liturgy of the universe. Every element in Christian belief and moral teaching, every aspect of our sacramental system, of our patterns of worship, and of our spirituality depend on the world about us. Indeed, the natural world is the primary revelation to us of the divine. Once we accept the universe as emergent reality, then what is said in Genesis, in the Psalms, in the writings of Saint John, in the Epistle to the Colossians, can be said of the universe as we now perceive it. This understanding expands rather than contracts our understanding of scriptural and theological statements.

In this emergent process we recognize transformation moments as those instants when the numinous presence in the phenomenal world manifests itself with special clarity. These moments include the time of a supernova explosion that produced the ninety-some elements needed for the solar system, for planet Earth, for life and consciousness. These transformation events can be considered cosmological moments of grace. Just as there are historical moments of grace and sacramental moments of grace, so, too, there are such cosmological moments. These are times that deserve ritual commemoration, just as moments in the annual solar cycle do; solstice and spring equinox are given ritual expression in nativity and resurrection liturgies. In most religious traditions, diurnal moments of transformation, dawn and dusk, the mysterious transition from day to night and from night to day, are profoundly religious moments to be observed with appropriate prayer and ritual.

Finally, I would like to note that I have emphasized the historical basis of Christian alienation because of concern over the ecological devastation of Earth. Only history can reveal how deep-rooted this alienation is within the Western psyche. The situation before us will not respond easily to any proposed remedy. The forces of destruction are too pervasive, the consequences of extensive change too painful. Any response will involve three aspects of the problem: principles, strategy, and tactics. At this time Christians have begun to respond in all three categories. My own concern is that we see the issue in its full depth and that the solutions offered be comprehensive enough and profound enough to save the biosystems of this late Cenozoic period, for these next few years are of crucial and immense importance. I call this our "great work."

We are in the process of losing the beauty and wonder of a gracious world designed in some manner as the place where the meeting of the divine and the human can be achieved in its full expression, a place suitable for the divine indwelling. Will we have the energy and the will to restore this world?

8.

Women Religious

Voices of Earth

(1994)

In these past two centuries an immense amount of work has been carried on by the various religious congregations of women throughout the world in the areas of education, nursing, social services, the alleviation of poverty, and spiritual guidance. There is no way of expressing the full extent of the work they have accomplished or the influence they have exerted.

Just now, however, women's religious communities—along with every other component of human society—are called to accept a new role, the most difficult role that any of us has been asked to fulfill, that of stopping the devastation that humans, principally those in our commercially driven societies, are inflicting on the planet. Otherwise the natural world will not survive in any integral manner, nor, in this situation will humans or Christians survive in any acceptable mode of fulfillment.

Survival, however, is not sufficient. A new type of community, a much larger Earth community, including all the living and non-living components of the planet, must now be brought into being. This would be a comprehensive community where humans would be present to Earth in a mutually enhancing manner. This has not been a primary human objective since we attained empirical knowledge of the structure and functioning of the planet.

The tragedy is that the extinction of species, the ruin of the forests, the loss of soils, the pollution of air and water, the thinning of the ozone layer—all these things are coming about through global economic drives shaped by the dominant influence of a Western civilization formed within a biblical-Christian as well as a classical-humanist matrix. The economic rapaciousness is not a trivial fault or a minor error in our thinking and acting. It is among the most devastating things that has happened to the planet since its emergence into being more than four billion years ago. It is the consequence of a profound failure deep in the religious-cultural tradition itself. Certainly it is as decisive a transition moment in the course of human affairs as was the beginning of the Neolithic period some twelve thousand years ago.

The transformation that will be needed is of an order of magnitude greater than any previous transformation of Western civilization, vaster certainly than the transition from the classical to the medieval period, or from the medieval period to the Renaissance, or from the Renaissance to the Enlightenment, or from the Enlightenment to our modern world. It is also much greater than any of the transformations that have taken place in other cultural and religious traditions.

Quite simply stated, we are terminating the Cenozoic period of Earth history. This term, "Cenozoic," refers to the last 65-million-year period when Earth and all its living forms as we know them came into being—the trees and grasses and flowers, the birds and fish and mammals particularly. In this situation every human being is involved, whether in formal religious life or not, whether men or women, from whatever region of the world, and from whatever profession or occupation. All our efforts are needed to bring to its full expression the emerging Ecozoic Era, the coming period when humans will be present to Earth in a mutually enhancing manner.

Humans in this period will regain sensitivities needed for the benign insertion of their activities into the ever-renewing processes of nature, or else humans will enter into a degraded mode of being. Our only security lies in an integral human relation with the life systems of the planet. Every human activity, every professional role, every religious tradition, must now be judged by the extent to which it inhibits, ignores, or fosters this mutually enhancing human-Earth relationship. We need an Ecozoic economics, an Ecozoic jurisprudence, an Ecozoic education, an Ecozoic medical practice. So, too, our religious and spiritual development lies in the transition from our present Cenozoic religious and spiritual life to life in an Ecozoic context.

What is preventing such a transition is that our present Western life orientation is so totally human-dominated in its values. This perspective is true of our religious lives as well as of our economic lives. It is a perspective in which the purpose of creation is seen as simply for human use, with everything else considered as background. Where did this human-dominated perspective arise? When the biblical world, at the time of the Exodus, began to perceive the divine primarily through the historical human order rather than through the cosmological order, this outlook made it possible for us to neglect the natural world in our emphasis on our meeting with the divine in the events of human life. Whatever the gains in the mode of human consciousness attained at this time, there was a danger in the changes it brought about. This danger of moving into an exaggerated anthropocentrism was vastly increased by Christian integration with the classical-humanist tradition of the Mediterranean world.

These historical developments made possible a single-minded concern for the human to the neglect of Earth so that we would in these later centuries abuse and plunder the life-sustaining functioning of the planet. We were not really concerned that the loss each year of some twenty-five billion tons of topsoil throughout the world would make it increasingly difficult and even impossible to nourish the increasing numbers of humans. Nor were we concerned that, after enabling people to survive, they would have an acceptable world in which to live—a world with the grandeur in its natural forms that would provide the psychic-spiritual as well as the physical needs of humans.

This is because as Christians we have been primarily concerned with divine-human and inter-human affairs in accord with the two great commandments, love of God and love of neighbor. Thus we have fulfilled the precepts of the law and the prophets. However effective this presentation of Christian spiritual teaching might have been in the past, it is no longer sufficiently comprehensive for Christian survival. There is a third component that cannot be neglected, namely, love of the natural world without which the human world cannot function in any effective manner. Earth entire was born of divine love and will survive only through our human and Christian love. Christians are ineffective just now largely because we have not understood the need of compassion for suffering Earth, the compassion expressed by Saint Paul in his reference to the world "groaning for deliverance."

At the present time the protest of the pillage of Earth, compassion for Earth, and commitment to the preservation of Earth are left mainly to secular environmental organizations as though the matter were too peripheral to be of concern to Christians. While recent statements made by Church authorities on environmental issues have been admirable, they have not always displayed the depth of understanding needed. Nor have they yet provided the inspiration required for an effective Christian movement for the survival of Earth in all its living forms of expression. So, too, our Catholic universities have not attained any distinction in the field of ecology. They contain a few courses on environmental thought, but these offer little real understanding, insight, or guidance for effective action. Nor have our religious orders yet shown significant leadership on ecological issues.

The Ecozoic Era and Christian Leadership

What then are the necessary conditions for moving into the Ecozoic era? The first condition is to recognize the human community as integral in its structure and functioning with the larger universe in which we live. There is no human community in any manner separate from the larger community of all those living and non-living beings surrounding us and on which we totally depend for every breath that we breathe. The created world forms a single community with Earth itself. So intimate is the human with every other being on the planet that we must say that the human community and the natural world will go into the future as a single sacred community or neither will survive in any acceptable manner.

Although life on Earth and the human mode of being may not be in immediate danger of becoming extinct, Earth and all its basic life systems are being severely and irreversibly damaged as the immense diversity of life is diminished. Even now we are living amid the ruined infrastructures of the industrial world and of the natural world itself—with all the physical, aesthetic, emotional, and spiritual implications of life in that condition. These ruins are becoming even more pervasive as the industrial world increases its exploitive control over the planet.

The second condition for effective action in the future is to recognize that planet Earth will never again function in the manner it has functioned in the past. Planet Earth emerged into being in all its grandeur over 4.5 billion years completely apart from any human decision, since *Homo sapiens* arose only some 150,000 years ago. Before the Industrial Revolution of the past few centuries, humans were not capable of damaging the planet to the degree that we are now able to do.

In the future, however, almost nothing will happen without human involvement. We cannot make a blade of grass, but in the future there is liable not to be a blade of grass unless we accept it, protect it, and foster it. In the future these will be the three basic human functions with regard to the natural world: to accept, to protect, and to foster. Through our science and our technologies we have intruded so extensively into the natural world that we cannot simply withdraw from the situation we have created. For example, until recently we did not need to think about the ozone layer. Now we must worry about it and take measures to protect it.

The question then arises concerning the role of Christians and especially the role of women's religious communities in creating the conditions for the emergence of the Ecozoic era. In answering this question we might reflect on the fact that over the centuries after the decline of the Roman Empire Christian religious communities were among the principal forces in creating an entire civilization, a civilization that we speak of now as the medieval world. It was an enormous effort, and it constituted one of the most creative periods in human history.

We are called to such an effort at the present time. Yet in its magnitude and form our task is far greater. Our destiny is to work with the larger Earth community in shaping an integral planetary process. This is not simply a Christian process or simply a human process, but rather a comprehensive process of planetary renewal in its geological and biological as well as in its anthropological aspects.

Voices of Earth

The task of renewing Earth belongs to Earth, as the renewal of any organism takes place from within. Yet we humans have our own special role, a leading role in the renewal, just as we had the dominant role in the devastation. We can fulfill this role, however, only if we move our basic life orientation from a dominant anthropocentrism to a dominant ecocentrism. In effecting this change we need to listen to the voices of Earth and its multitude of living and non-living modes of expression.

We should be listening to the stars in the heavens and the sun and the moon, to the mountains and the plains, to the forests and rivers and seas that surround us, to the meadows and the flowering grasses, to the songbirds and the insects and to their music especially in the evening and the early hours of the night. We need to experience, to feel, and to see these myriad creatures all caught up in the celebration of life. We especially need to hear the creatures of Earth before it is too late, before their voices are stilled forever through extinctions occurring at such a rapid rate. Once gone they will never be heard again. Extinction is forever. The divine experience they communicate will never again be available to humans. A dimension of the human soul will never be activated as it might have been. None of the wonders of the human can replace what we are losing. However, to speak of the voices of the natural world is to become suspect to some "religious" people, for the Western religious traditions have developed a suspicion of such attitudes toward nature, devaluing them as simply "pagan" or "animistic" notions. We have lost sight of the fact that these myriad creatures are revelations of the divine and inspirations to our spiritual life.

Our inner spiritual world cannot be activated without experience of the outer world of wonder for the mind, beauty for the imagination, and intimacy for the emotions. If we lived on the moon our minds would be retarded in their development, our imagination would be as empty as the moon; our emotions would be as dull. Our sense of the divine would reflect the desolation of the lunar landscape. Our Christian spiritual life is already being diminished as the basic faculties of the human soul are denied their inspiration from the larger context in which they function. The ultimate reason that we cannot live on the moon or on some other planet such as Mars is that our souls would shrivel into their unfulfilled selves.

Through our contact with the natural world we learn that the universe throughout its vast extent in space and throughout its long sequence of transformations in time is a single multiform celebratory event. Our role is to enter into this celebration in a special mode of conscious self-awareness, for this celebration is the divine liturgy, the purpose of all existence, a celebration begun in time but continued through eternity.

To save the integrity of this celebration is the first lesson in survival, for this is the context of all the productivity of the planet as well as our primary experience of the divine. If we fail to enter into this celebration, if we seek simply to exploit the myriad creatures about us, then they will fail to produce their fruits and the grand cycle of existence will be diminished. This indeed is already happening.

We, and our children, are becoming autistic in relation to the natural world. We live in a world of computers, cell phones, digital photography, television, highways and automobiles, supermarkets, and trivial plastic playthings for our children—all fostered by inescapable advertising aimed at stirring our deepest compulsions to buy and consume. Our education is focused on producing skills associated with the production, distribution, and use of such a multitude of objects with none of the exaltation of soul provided by our experience of natural phenomena. We no longer realize that the universe is a communion of subjects, not a collection of objects—subjects to be communed with as divine manifestation, not objects to be exploited solely for economic gain.

The Role of Women Religious Historically and at Present

If in past centuries women religious have been dedicated to educating, healing, and guiding the human community, the primary role of religious congregations of women at present might well be to preserve Earth from further devastation. This work is a condition for fulfilling any other role. The natural and human imperatives are prior to and a necessary condition for any Christian imperative. We cannot be integral Christians because we are not integral humans. We are not integral humans because we have alienated ourselves from the larger life community.

This reorientation of Christian thought and action from its completely human preoccupations to a concern for the larger sacred community lies in the sequence of changes that began in the seventeenth century when Francis de Sales established the first modern laypersons' spiritual guide with publication of his *Introduction to a Devout Life*. Inspired by his work, Vincent de Paul founded the Daughters of Charity in 1634. Shortly after this a group of six women in Le Puy, France, founded the Sisters of Saint Joseph in 1650 with the assistance of Jean Pierre Medaille, SJ. These initiatives involved monumental changes in the religious role of women in the Church beyond the convent because the Council of Trent in the prior century had defined religious life for women as having to be lived in cloistered enclosure. Further changes took place in the nineteenth and twentieth centuries with the founding of a the long list of women's religious congregations for teaching, care of the poor and afflicted, missionary work, and a multitude of other specialized concerns. In 1809, Elizabeth Ann Seton founded the Sisters of Charity in this country for the education of children and for works of mercy. Katharine Mary Drexel founded the Sisters of the Blessed Sacrament for missions to African Americans and Native Americans.

What is clear is that until our own century no religious community has been founded to protect Earth from devastation increasingly inflicted on the natural world through human agency. Such would in former times have been unthinkable. Now, however, there is hardly any religious or human activity that has prior claim to our concern, since everything else becomes irrelevant if the natural world undergoes further devastation. If a women's religious congregation committed to the saving of the natural world was unthinkable in former centuries, it is now unthinkable that any such congregation should not be committed to this task. If the life systems are not saved, then everything else is irrelevant. None of the other purposes whatever can now be fulfilled except in this ecological and Ecozoic context.

A new awakening is indeed occurring in women's religious communities. Much of this development has taken place in an agricultural context, since the first step in human-Earth relations is to associate our presence to Earth in its life-giving wonders. In 1980 Sister Miriam Thérèse MacGillis, of the Dominican Order, a leader in this movement, founded Genesis Farm in the Delaware River watershed in Blairstown, New Jersey, as an Earth-literacy center. It has brought forth a community-supported agricultural project and has initiated and supported other regional projects in many aspects of sustainability. Sister Miriam Thérèse MacGillis has also developed an accredited academic program exploring the universe story and its implications for human meaning and social transformation.

A long list of other religious women could be mentioned in this context. Sister Paula Gonzalez in Cincinnati has shown how we can diminish our dependence on fossil fuels by a more profound understanding of the sun and how it can heat our buildings as well as inspire us spiritually. Sister Mary Ann Garisto, a Sister of Charity of New York, has established an impressive agricultural project upriver from the city. It connects people and neighborhoods from the city with food cultivated on the farm. Sister Rita Wieken has done similar work on the Franciscan lands in Tiffin, Ohio. On the motherhouse lands of the Dominican sisters of Amityville, New York, a project initiated by Sister Jean Clark is spreading Earth-literacy in communities on Long Island. Sister Virginia Pearl, a Sister of Saint Joseph, is working with Heartland Farm outside Salina, Kansas. Sister Chris Loughlin has established an Earth-literacy center called Crystal Springs south of Boston. Then there is the new venture in Vermont to found a monastery for women led by Gail Worcelo and Bernadette Bostwick. Named Green Mountain Monastery, it is dedicated to protecting and preserving Earth by modeling ways of supporting a new pattern of human presence to the planet. Of special note is a remarkable project initiated by Dominican sister Patricia Siemen who, with the support of her order, has created a Center for Earth Jurisprudence jointly sponsored by the law schools of St. Thomas University and Barry University in Miami, Florida.

In addition there are the extensive properties owned by religious communities, properties that are still in a relatively undisturbed state, where wildlife is often abundant, where human predation is limited, where the primordial impress of the divine can still be felt, and where a sense of the sacred is available. Preservation of such lands is one of the great urgencies of the moment. Religious orders have been so absorbed in the salvation processes of the human that they have had little time for recognition of the profound need of the human soul for contact with natural processes. Now, however, we might recognize that these regions where wildflowers grow and where birds nest have become infinitely valuable as places more needed than ever by the human spirit. Such regions might be thought of as shrines where the pristine impress of the divine can still be experienced. To lose these lands to development would be an irreparable spiritual loss.

The appeal of this larger dimension of Christian concern for Earth has led to a remarkable movement, namely, the founding of an association called "Sisters of Earth" made up of both nuns and laywomen. The group has been meeting every two years for the past ten years. This network of women from diverse religious communities and beyond is a significant venture, both for the movement of general concern for the natural world and for the religious life within its Christian context. All of these women have discovered the larger dimension of the universe and Earth as the context for their work and their spiritual practice. Many of the activities of these women are described in the book by Sarah Taylor titled *Green Sisters*, published by Harvard University Press.

The Primacy of Human-Earth Relations

To propose that human-Earth relations have a claim on our attention prior to divine-human or inter-human relations is simply to assert that our experience of the universe is prior to our experience of ourselves or of the divine. As Saint Paul mentions in the first chapter of his Epistle to the Romans, from the things that are made we come to know the Maker. These three—divine, natural, human—are so integrally connected with each other that none can function effectively without the others. Yet, because of an over-balance in favor of divine-human and inter-human relations, the sense of the integrity of the divine, the natural, and the human has been neglected for centuries by Christians.

In the past Christians have been concerned with internal religious conflicts over problems of salvation, biblical interpretation, theological explanation, sacramental administration, and ministerial functioning on the one hand, and concern for human well-being on the other. These concerns and the pervasive ambivalence of Christians toward the natural world, especially since the sixteenth century, have produced insensitivity to natural life systems. Even Saint Thomas thought that the natural world was for the glory of God and the use of humans. It had no rights in itself. A persistent effort in Christian spirituality has been to distance ourselves from the natural world, to become detached from rather than committed to our surroundings. There are exceptions to this, as when the natural world did come into focus through such individuals as Hildegard of Bingen (1098–1179) and Francis of Assisi (1181–1226).

After the sixteenth century, especially after modern science— with its abandonment of a personal creative deity as origin and ruler of the universe—began to develop, Western Christians became further alienated from the world of natural phenomena. When science began teaching that humans had originated from within the processes of the natural world, this view led to a negative reaction and further Christian emphasis on the human as a spiritual being in a non-spiritual natural world.

Now this opposition between Christianity and science is largely in the past. We know at this time that the universe story and the human story constitute a single story. There is no human without the universe, no universe without the human. Devastation of the outer world is simultaneously devastation of the inner world. To be isolated from the phenomenal order of the natural world is to be alienated from the deeper dimensions of our own being.

The responsibility is universal, and goes beyond ethnic or gender or occupational identity, beyond clerical or non-clerical status in the Church. Yet each of these has its own special role. So, too, there is a special role for women in this situation, a role that has come to take on increasing importance. The eco-feminist movement joins some of the most powerful movements of our times in effecting the transition from a non-viable to a viable mode of existence for the planetary community, namely, the transition from the terminal Cenozoic to the emerging Ecozoic era in Earth history. It is a matter of following the guidance that nature gives to herself in ordering the affairs of the planet.

It might well be that women are attuned to the voices of Earth in a way especially needed as we move into a future less dominated by the plundering processes of the industrial nations. Women may help us recover our reverence for natural manifestations of the divine in the world around us. The time has come when the single greatest service that women religious can make to the larger destinies of the human, the Christian, and the Earth community is the recovery of our human and Christian intimacy with all those wonderful participants in the universe of being.

Acceptance of such intimacy with the Earth community is a condition for transcending the mechanistic life attitude that dominates the modern world. It is also a condition for entering into the poetry of existence, for composing music, for creating the visual arts. In all these areas the inner form of things is experienced in and through their physical reality. This inner form is the voice listened for and heard when we become present to Earth in some significant manner. Only if Earth is perceived as precious, only if there is a profound intimacy with the inner self, the inner spontaneities of nature, will we have the will or the psychic energy to bring about the difficult transformation in human life required of us. We are nourished by the natural world both physically and spiritually. We cannot have the physical without the spiritual.

For the foreseeable future, this unity of humans with their environment will almost certainly be a primary context for inspiring the various activities of women's religious communities. Whether their work is teaching or healing or community building or fostering movements for social justice, or educating, or more direct spiritual or religious activity, none of these will any longer succeed apart from the larger context of the natural world. This larger world is the only integral community of existence. It is the only effective context for physical survival, for healing, for religious ritual or spiritual development. Religion itself is awakened in the human soul by our experience of the awesome qualities of the immense universe about us, its overwhelming grandeur, its terrifying as well as its entrancing qualities. As the grandeur of the natural world declines, the primordial manifestation of the divine is progressively diminished.

The Universe Story

All of this must find its expression in the story of the universe. Indeed, the various civilizations of the world are generally founded in some story indicating how things came into being in the beginning, how they came to be as they are, and role of the human in the story. This story of the universe is eventually the context of education, of healing, and of any other activity in which humans engage.

This story is the story that the universe tells of itself. It is the story told by every being in the universe, by the stars in the heavens, by the mountains and rivers of Earth, by every wind that blows, by every snowflake that falls, by every leaf in the forest. To know this story of the universe as our sacred story is to have an adequate foundation for the task before us. This story tells us who we are and how we came to be here and what our lives are all about.

For the Christian it tells us of the Trinity in the three most basic tendencies of the universe: differentiation, interiority, and universal bonding. These deepest tendencies of the universe, which manifest the ultimate divine forces that brought the world into being, can provide us with a profound way of thinking about Father, Son, and Holy Spirit. So, too, the universe story can be told as the Christ story in accord with the teaching of so many Church Fathers as well as theologians and spiritual teachers over the centuries.

Thus the voices that resound throughout the universe are ultimately the divine voice resounding in the immense variety of its modes of expression. Precisely in this immense differentiation of natural phenomena lies, in the phrase of Saint Thomas, "the perfection of the universe." In this perfection the divine order of existence finds its most complete manifestation. At the end of his *Divine Comedy* Dante tells us that in his vision of divine reality he saw "all the scattered leaves of the universe bound by love in one volume." Such is the origin and end of all our human or Christian or religious communities.

9.

The Wisdom of the Cross

(1994)

My subject here is the wisdom of the cross in relation to our new understanding of the universe. We have come a long way from the classical worlds of Genesis, Plato, Aristotle, and Ptolemy, from the medieval worlds of Francis of Assisi and Thomas Aquinas, even from the early modern worlds of Copernicus, Descartes, Galileo, and Newton. We are into the quantum and expansive worlds of Einstein, Planck, and Hubble. Beyond these we have entered the mysterious world that comes to us from the background radiation observed by Wilson and Penzias at Bell laboratories in New Jersey during the 1960s. These efforts have given us a New Story of the universe, a story that we did not have in previous centuries when we experienced the universe as a sequence of seasonal transformations.

What is significant about these recent views of the universe is that they are more extensively grounded in empirical observation and rendered in numbered figures, interpreted as hypotheses or theories. There is still a "mythic" dimension to our ways of understanding the universe, even though the foundations of our understanding rest more deeply in the precision of our empirical observation. These new empirical ways of viewing the universe differ from the earlier views mainly because of the lack of spiritual or religious meaning that we find in them. We have grown so accustomed to viewing the natural world in terms of scientific equations and economic values that we hardly realize how indifferent we have become to the spiritual dimension of the universe. This can be attributed in part to our emphasis on the scriptures, and to personal interior experience as the exclusive source of our religious experience. It is also due to the random, mechanistic interpretation that scientists have given to their discoveries.

The simple truth is that most of us no longer live in a sacred universe. This lack of a sacred context of existence causes us to feel alienated from and even antagonistic toward the natural world. We experience the world about us as a collection of "natural resources" for our economic exploitation, or simply as the surrounding "environment" of the human. Our experience of the universe as the presentation that divine reality makes of itself to human understanding is almost completely lost. Even when we view the distant mountains or look out over the oceans or see the stars at night, what we experience is a momentary exaltation of spirit, not the sense of a pervasive divine presence with decisive control over our lives.

My suggestion is that we take a serious look at the universe as this reveals itself to us in this century. We now experience the universe, within the phenomenal order, as a self-emergent process that has gone through a long sequence of transformations through the millennia to become the world about us. This sequence of transformations is vastly different from the ever-renewing seasonal transformations of the universe as it was known in prior centuries, for this emergent process is irreversible, while the universe as formerly known was seen as a sequence of ever-renewing seasonal cycles.

Discoveries of how the universe, the planet Earth, and we ourselves have come into being have so challenged our Christian understanding that we are still unable, intellectually or emotionally, to feel fully at home in this context. We have little appreciation even of the planet Earth that is unique in our solar system in its ability to bring forth its astounding variety of living beings. In this it is unique among all the planets of our solar system.

We ourselves were brought into being through this process. The universe story and the human story are a single story. We are so intimately associated with the world of the living that we must consider ourselves as cousins to every other living being. Yet, because we are so alienated from the universe in its unfolding reality, we do not appreciate our place or our role in this process. We do not live in a universe at all. We live and function in a cultural tradition, in an economic order, in a world of political allegiance, not in a physical universe or in what we generally refer to as the natural world.

Many people in religious settings who live more intimately with the land have been taught that it is spiritually preferable to keep apart from attachment to this world. Thus in the Advent season we pray that we may "love the things of heaven and judge wisely of the things of Earth." To judge wisely of the things of Earth means to keep a detached stance in order to avoid being emotionally bonded to the magnificent and "seductive" aspects of the planet. During other seasons we separate out the heavenly world from the earthly world in such a way that this alienation is deepened. The presence of the divine in the natural world is obscured or diminished in our consciousness. Alienation deepens into suspicion and antagonism.

Even when we live in a spiritual universe, it is the ancient universe of our religious imaginations. This is the universe in which our religious traditions took shape, not the universe as we experience it at present. Our alienation from the natural world is paradoxical because the universe in all its grandeur and awesome qualities has generally been considered the primary manifestation of the divine, not only by Christians but also by many other peoples of the world. For example, among the world's indigenous peoples, where coherent communities are still intact, we find that religious imagination in which intimacy with local bioregions and biodiversity frame the sacred in a seamless continuity with all of life's demands. These religious ways are ancient in human social formation. Early stories and myths served to shape humans in creating dynamic life ways in relation to natural processes. Our modern alienation appears to have drawn us away from this deep instinctual patterning with Earth itself. How could this have happened? What intervening forces have pushed aside the wisdom of Earth?

For most Christians the difficulty is that the biblical story is a narrative of human-divine interaction that takes place against the background of a universe that is fixed in its abiding sequence of seasonal renewal. The universe itself has no story except that of its original formation. Afterwards it is a fixed stage upon which the human-divine drama is enacted. This we can observe in the *City of God* of Saint Augustine. In this work he tells the story of the world and the divine design for the world. But after the early treatment of creation the entire story is the human story.

While the emergent universe should be experienced as entirely compatible with Christian belief, it has until now been experienced as alien, ancillary, and even contradictory to authentic doctrine. Nowhere do we hear our New Story of the universe told as sacred story. This I propose is a central issue involved in the discussion that is before us. How does the wisdom of the cross function in this new context?

The more we understand the universe story as this is now available to us, the more clearly we see that the wisdom of the cross and the wisdom of the universe are two aspects of a single wisdom, that the universe and the cross are integral parts of a single story. Neither is complete without the other. The order of the cross is coherent with the order of the universe. There is ultimately a single wisdom that reaches from end to end mightily and orders all things sweetly.

Redemptive wisdom cannot be alien to creative wisdom. There has been no mistake. When we speak of the wisdom of the cross being adverse to the wisdom of this world, we must understand this as referring to a false wisdom, not to the true wisdom of the universe. This type of insight is not new in Christianity. Indeed, medieval thinkers presented a range of perspectives on cosmology and creation that attempted to integrate their understanding of the wisdom of the cross in relation to the wisdom of the world.

It is helpful to recall that Christian medieval scholastic thinkers presented a wide range of cosmological thought and that even prior to the medieval age work was done in this area. For example, John Scotus Eriugena (c. 800–877), a Carolingian thinker, was a Neo-Platonist who lived prior to the reintroduction

of Aristotelian thought and its emphasis on form in the material world. Yet, his knowledge of the Greek Fathers, especially the
Cappadocians, Pseudo-Dionysius, and Gregory of Nyssa,
brought him to a sense of a dynamic outgoing and return of creation to the One God. Thus, the divinization process, or *theosis*,
so evident in the Greek Christian writings took on a cosmological character for Scotus Eriugena. In this sense he had a pen-
chant for the Dionysian saying from the *Celestial Hierarchy*,
"for the being of all things is the Divinity above being."[1]
Through this line of thinking Scotus Eriugena came to see creation as a divine outpouring, or manifestation, of the divine. He
wrote: "It follows that we ought not to understand God and the
creature as two things distinct from one another, but as one and
the same. For both the creature, by subsisting, is in God; and
God, by manifesting himself, in a marvelous and ineffable manner creates himself in the creature."[2]

In his thirteenth-century writings, Thomas Aquinas was
strongly influenced by the reintroduction of Aristotelian thought
into the medieval Christian world. One of Thomas's creative
insights was the determination of any analysis as starting from the
generalities in the physical world. This straightforward observation brought him to the realization that the human search for ultimate or first causes begins with a wisdom embedded in human
knowing the things of the world. For Thomas, the divine can be
known only indirectly through the effects of God's creation in the
world. What appears in human action, then, is a search for this
good in creation. Thomas repeats many times that the greatest
good of the universe is the order of the universe imparted through
the three persons of the Trinity. Thomas writes that:

> . . . in all creatures there is found the trace of the Trinity,
> inasmuch as in every creature are found some things

[1]*Celestial Hierarchy* IV, 1, and III, 686d; Patrologia Graeca III,
177d1–2: *to gar einai panton estin he hyper to einai theotes.*

[2]*Periphyseon*, III. 678c.

which are necessarily reduced to the divine Persons as to their cause. For every creature subsists in its own being, and has a form, whereby it is determined to a species, and has relation to something else. Therefore as it is a created substance, it represents the cause and principle; and so in that manner it shows the Person of the Father, Who is the "principle from no principle." According as it has a form and species, it represents the Word as the form of the thing made by art is from the conception of the craftsman. According as it has relation of order, it represents the Holy Ghost, inasmuch as He is love, because the order of the effect to something else is from the will of the Creator.[3]

In Thomas Aquinas we find an orientation toward the world as manifesting the Trinity, the order of the universe.

Roger Bacon in the mid-thirteenth century drew on Aristotelian thought to describe the order of the universe evident in created reality in line with Thomas Aquinas. He wrote: "This follows because a universal is nothing other than a nature in which singulars of the same [nature] agree; but particulars agree in this manner in a common nature predicable of them, without any act of the mind."[4] Bacon posits that this inherent order of the universe in the human, which is even deeper than the mind, provides a wisdom orientation through the world toward the divine.

John Duns Scotus (1266–1308) likewise argued, as did Thomas, that our knowledge of God begins with our encounter with creatures. He differed from Thomas in that he thought that our knowledge was univocal and ultimately descriptive of reality in contrast to Thomas's understanding of the analogical character of human knowing and naming of creatures. In this

---

[3]Thomas Aquinas, *Summa Theologica* (I, q. 45, a. 7).

[4]*Opera hactenus inedita* (Vol. I–XVI), ed. Robert Steele (Oxford, 1909–1940) VII:242–43.

brief consideration of medieval Christian thinkers on cosmology there is a range of views but they all relate the wisdom of the cross to an inherent wisdom in the world.

This being so, we might expect that for contemporary Christians, who have inherited the richness of these thought traditions on cosmology, the universe itself would reveal a wisdom related to, and indispensable to, the wisdom of the cross. But we find that instead of building on these older traditions that saw in creation the most immediate revelation of the divine, a disdain of creation has appeared in contemporary Christianity. To disdain the wisdom revealed in the universe in order to exalt the wisdom of the cross might appear admirable from the standpoint of religious fervor, but it has been problematic, especially when taken literally. Such a position can result only from a distorted view of the integral revelation that the divine makes of itself.

In such a distorted context the wisdom of the cross constitutes, not exactly our hope, but the source of our difficulty, for once we believe that the wisdom of this world is foolishness, then the world about us deserves only the disdain that we impose upon it. In this context the planet Earth would rightly be conceived as existing simply as background for the human and we would pride ourselves on the extent of our alienation from the universe in its basic structure and functioning. Nevertheless, we would be left with the question of how we are to relate to the revelation of the divine that is presented to us in the universe.

This brings us to those passages in scripture that speak of the Universe-Christ. We find this, of course, in the Epistle to the Colossians where Saint Paul indicates that in Christ "all things hold together" (Colossians 1:17). We might expect a certain coherence between the grandeur of the universe and the majesty of the cross of Christ. For the Christian world, these remain the reference points as regards reality and value. The mystery of the cross can be matched by the mystery of creation. Neither is within human comprehension. The New Testament is filled with references to the ongoing process whereby the Gospel brings about its fulfillment in the formation of the kingdom of God. So also in the Epistles of Saint Paul we find references to history as filling up those things wanting in the sufferings of Christ. It would seem quite appropriate that in a later period the emerging universe process and the emerging redemption process should be brought together into a single coherent process.

This coordination can be understood quite clearly when we consider the central role of sacrifice in the redemption process and then observe the central role of sacrifice in the unfolding of an emergent universe. We might even say that the redemptive suffering of Christ lies in the line of creative transformation moments revealed to us in the universe throughout the entire course of its history. This is most obvious in the transformations experienced during the decisive moments in the shaping of the universe as it moved from lesser to greater complexity and consciousness. These moments have a sacrificial aspect.

Such a cosmological moment occurred, for example, when a supernova exploded in enormous heat, scattering itself as stardust out into the vast realms of space. In the heat of this explosion the ninety-some elements were formed. Only then could the planet Earth take shape, life be evoked, and reflective intelligence become possible. This supernova event could be considered a sacrificial moment, a cosmological moment of grace that established the possibilities of the entire future of the solar system, Earth, and every form of life that would ever appear on Earth, including the spiritual dimension of the human mode of being.

Such cosmological moments of grace are necessary before there can be historical moments of grace or Christian moments of grace. All three can be considered religious moments of grace. All are guided by what might be considered an inner wisdom determining the course of the universe, for the universe and the emerging process of its unfolding is, from its primordial emergence, a spiritual as well as a physical process. Such transformation moments necessarily have a catastrophic aspect. Indeed, these cosmic moments of reshaping occur amid awesome violence. The world is born into a radically new phase of its existence. The cosmic recasting of the universe and of the human order of being by the redemptive sacrifice of Christ was preceded by these earlier moments. They are not unrelated events.

In the history of the Earth a creative moment occurred when newborn cellular life was imperiled by the presence of free oxygen in the atmosphere. The earlier life forms that produced oxygen could not themselves live in contact with oxygen, for while living beings as we know them cannot do without oxygen in proper amounts, free oxygen was originally a terrible threat to every living form. It was a threat even to the rocks as these were oxidized.

For a proper balance to be achieved and then stabilized a moment of grace had to occur, a moment when some living cell would invent a way of utilizing oxygen in the presence of sunlight to foster a new type of metabolic process. Photosynthesis was completed by respiration. At this moment, under threat of extinction, the living world as we know it began to flourish until it shaped Earth anew. Daisies in the meadows, the song of the mockingbird, the movement of dolphins through the sea—all these became possible at this moment. We ourselves became possible, along with music, poetry, and painting.

In a special manner religion became possible. The incarnation and all the mysteries associated with Christian revelation emerged. All these had their deepest origins in the transformation moments that had not only a physical-material aspect but also a psychic-spiritual aspect. If we are to take Saint Paul and Saint John seriously, then there was also a primordial numinous or Christ aspect to such events.

If we can accept this Christ-Universe equation, these must be seen in terms of each other. This is especially true with regard to the sacrificial dimension through which each achieves its purpose. This sacrificial dimension is a scandal in both instances. Within the functioning of the universe in its daily survival as well as in its continuing unfolding, there is a sacrificing of each reality for the others. This dying that others might live is something of a universal experience. It is also another phase of the mystery of the cross, the mystery of creative immolation.

Once the human enters onto the scene, there is a further determination to be made by conscious human decision concerning the destiny of the universe, but especially of the planet

Earth. Here we come to the basic concern of our discussion, the ability of humans to enter consciously into both the creative processes of the natural world and into the redemptive processes of the Christian world. We are, it seems, constantly tempted to turn the redemptive program against the universe program.

Our difficulty in appreciating the larger community of existence, which includes the non-human as well as the human, can be seen in our refusal to grant any moral or legal rights to the natural world. We accept the view of theologians generally, at least the classical scholastic theologians as well as the modern Kantian theologians, that any rights of the non-human world are not recognition of rights inherent in the non-human mode of being but are simply obligations that we owe to ourselves. Indeed the natural world, we say, is there for human use; the natural world does not constitute with the human a single integral society with every member of this society having its own proper rights to be what it is and to fulfill its purpose in the order of the universe.

Obviously "rights" is not a univocal but an analogous concept. Each being has rights according to its mode of being. Trees have tree rights, birds have bird rights, and humans have a primary obligation to respect these rights that reside in the non-human order. This is quite different from saying that any obligation to the natural world is simply an obligation to ourselves. Because of the latter attitude toward the natural world, we are plundering the planet in such a way that little remains that gives primordial witness to the divine revelation that comes to us through the wonder and majesty, awe and fear evoked by natural phenomena. This is so severe that we are now being told by competent biologists that we are killing our world and that the extinction of species taking place just now is unequalled since the collapse of the Mesozoic period some sixty-five million years ago. Not only are we devastating the biological systems of the planet, we are also changing the chemical balance of the planet in a deleterious manner. With our toxic poisoning of the planet we are in a few decades upsetting the balance that took the universe some hundreds of millions—even billions—of years to achieve. In America alone we are presently making two-hundred million tons of industrial chemicals each year in combinations mostly unknown to nature and that nature cannot absorb.

It is somewhat strange that Christians generally feel little responsibility for this situation. Since we are sublimely dedicated to spiritual purposes, it seems improper that we should concern ourselves with the fate of this world. We seem not to hear the divine voices sounding throughout the natural world, or the splendor of the divine revealed in natural phenomena. We seem not to realize the consequences of losing these manifestations of the divine. Our theologians somehow seem unmoved by the statements of Scotus Eriugena that God "creates himself in creatures," or of Teilhard de Chardin who spoke of a "communion with God through the Earth." We seem unaware of the profound dependence of our sense of the divine on our experience of natural phenomena. Nor apparently are we sensitive to the Christ-Universe equation and the Christ-Earth equation that we express each day at the Eucharist where the presider says of Christ that through Him and with Him and in Him are all things.

The difficulty here, as I indicated earlier, is that our intense attachment to the "spiritual" wisdom of the cross tends constantly to alienate us from divine wisdom as it is revealed in the universe itself. Presently we seem unable to deal with this twofold manifestation of divine wisdom. We are isolating the human from the natural systems of which we are a part. In the political, religious, economic, and intellectual establishments of the human, in the state, the Church, the universities, and the corporations we can observe the same difficulty, the radical discontinuity of the human from the non-human that we are imposing on the planet, and the consequent exploitation of the non-human by the human.

Obviously, when we read the scriptures or survey the various aspects of Christian theological traditions, we find a constant effort to maintain the unity of the comprehensive order of the universe within its divine matrix. Yet the difficulty is in the feeling of being in a state of exile from our true home as long as we are on this Earth. This deepens our sensitivity to the general pains of life that we identify as the "human condition." Our sensitivity is increased by several orders of magnitude when we are promised in the Book of Revelation that a time will come when we will be relieved of the human condition, a time when the Great Dragon will be chained up for a thousand years, a millennium. During this period Earth will be ruled over by the blessed and the great stress of battle will be diminished.

If this period of peace and justice designated as the millennium has been a source of hope for Christians, it has also been a source of deep disquiet, for it is a hope that remains unfulfilled in the course of historical time. The consequence has been an unease in the Western soul. In the Western psyche there is a deep hidden rage against the human condition based on a disappointment that the messianic expectation has not been achieved, that the millennium dream has not been fulfilled. This sensitivity toward the natural world was further increased by the experience of the plague that afflicted Europe from 1347 until 1349 when something like a third of the people of Europe died. Since the people of those times had no understanding of germs, they could only attribute the event to a moral cause. The world had become wicked and was being punished. The conclusion was to seek redemption out of the world rather than to look for more intimate relations with the world.

All these experiences and disappointed expectations are in the background of the Western drive toward dominion over Earth and the pathological plundering of its resources. The Christian tradition of a spiritual order over against the natural order of the universe has left the natural world defenseless against such an onslaught. The universe as divine manifestation has receded into the background of Christian consciousness. The sense of a Christ presence or a Christ identity with the natural world has been diminished.

The planet that demands a certain sacrificial aspect of all life is no longer acceptable. Because the millennial visions of Saint John in the Book of Revelation have not been fulfilled by divine grace, then humans have determined to fulfill these visions by human effort. The determination is to take this world apart and rebuild it completely, according to human design, if this is needed to achieve our "wonder world" expectations.

This extravagant expectation of a future built on human plundering of Earth's natural resources is largely responsible for the oppression of humans under our modern economic institutions. It is not only natural resources that are involved in this effort but also human resources. The consequence is an ever greater exploitation of the weaker by the stronger, of the less competent by the more competent, of those who own nothing by those who own everything. While such abuse has always occurred in human history, it has become especially critical in these times. We could endure the limitations of life and our human condition more graciously in a human community of shared benefits and burdens. We need only reflect on both the wisdom of the cross and the wisdom of the universe, namely, that sacrifice with each other and for each other is a dimension of life itself and should lead to an expansion rather than to a diminishment of life for each of us.

So too with ethnic differences. We have never learned the wisdom of the universe, a wisdom that teaches that difference is a primary condition for there even being a universe. It is also the reason why Saint Thomas speaks in the *Summa Theologica* (I, q. 47, a. 1) of difference as "the perfection of the universe." The racial and cultural diversity among the peoples of Earth is among the greatest splendors of Earth. Each ethnic group and each culture is enhanced rather than diminished by the manner in which the distinctive qualities of each group increase the grandeur of all the others. Finally, I would suggest that there is need for peace with Earth if there is to be peace among the peoples of Earth. Our alienation from Earth is one of the most significant causes of our alienation from each other.

We are now entering a new period in the religious-cultural history of our Christian-derived civilization. In this new period there is some doubt about the next generations being religious in the sense that prior generations have been religious according to Christian doctrinal or ethical imperatives. Nor can we expect that in a secularized society the coming generations will be religious out of any scriptural basis for their thinking. The sacred writings that guided us are unlikely to guide those so extensively alienated from the traditions that we have known. In such a situation there is a need for something beyond a sacred book that teaches the spiritual wisdom of the cross. What is needed, I suggest, is the additional wisdom of the universe. There is a need of sanctions other than the spiritual sanctions indicated by Christian belief. I suggest that there is a need for the wisdom of a universe that can impose physical sanctions upon whoever violates its laws, a universe that establishes the conditions of life itself.

The experience of our generation is a parable that future generations might well consider. Because we have violated the conditions of life by our assault on the planet, the planet is withdrawing the pure air and water and the fertility of the soil. The immense shoals of fish that once flourished in the oceans are no longer there. The exuberance of life that once existed on the planet can no longer be found. Nature cannot endure the afflictions we impose on it. We will obey the divine directions of the natural world or we will die. This is the ultimate imperative from which there is no escape.

Just what the future might be we cannot know with any clarity. We should at least be able to see that the wisdom of the cross presented in opposition to the wisdom of the universe cannot be an effective answer to the difficulties that we confront at the present time. We might also be able to see that our New Story of the universe should be understood as a sacred story in which divine mysteries are revealed to us with a clarity that we have never before known. Finally, we might want to consider that we are just now at a unique moment when the wisdom of the cross, in this new context, can arrive at a more expanded expression of itself.

10.

The Universe as Cosmic Liturgy

(2000)

Each morning we awaken as the sun rises and light spreads over Earth. We rise and go about our day's work. When evening comes and darkness spreads over Earth we cease our work and return to the quiet of home. We may linger awhile enjoying the evening with family or friends. Then we drift off into sleep and that dream realm beyond consciousness where our lives are renewed after the exhaustion of the day. As in this day-night sequence, so in seasonal sequence we experience changes in our ways of being. In autumn our children may spend their days in school and we alter our daily regime accordingly. In springtime we may go out more freely into the warmth of sunshine where some of us plant gardens. In summertime we may visit the seashore to find relief from the limitations that winter imposed upon us. In each of these seasons we celebrate festivals that give human expression to our sense of meaning in the universe and its sequence of transformations.

Our concern in this essay is to offer an overview of the many ways in which the integral dimension of the universe is manifest in different human communities throughout the Earth community. Just now the human community has a remarkable scientific understanding of the universe and of the planet Earth, yet, humans seem not to have the intimate rapport with the universe that earlier humans once had. Rather than fostering a mutually enhancing relationship with the other members of the Earth community, we have been responsible for causing dysfunction throughout the entire planet. Yet, along with our anthropogenic alterations of Earth's climate, soils, and waters are alternative visions of human-Earth relations, some of which are quite old while others appear dimly on the horizons of our future. Echoes of an ancient human-Earth relationship are evident among indigenous peoples, many of whose sustainable life ways have been significantly altered by contacts with modern industrial cultures.

The ethnography of the Omaha peoples, formerly living as a tribal group in the northern plains of North America, records a personal relationship with the larger universe that was ritually established at the time of each child's birth. Among these peoples a newborn infant was taken out under the sky and presented to the cosmos:

> Ye sun, moon, stars, all ye that move in the heavens,
> I bid you hear me! Into your midst has come a new life.
>   Ho!
> Consent ye, we implore, make its path smooth
> that it may reach the brow of the first hill.
> Ho! Ye winds, clouds, rain,
> All ye that move in the air
> I bid you hear me, into your midst has come a new life.
> Consent ye, we implore, make its path smooth
> that it may reach the brow of the second hill.

The invocation continues to address the hills, rivers, trees and all that lives on Earth with a corresponding request that the child be protected to reach the third hill. The birds that fly in the air, the animals great and small, that dwell in the forests, the insects that move among the grasses—all are invoked. Then a final petition asks that all creatures everywhere will take care of the child, that it may travel beyond the four hills. This legacy remains a significant expression of the Omaha sense of the cosmos as liturgical presence.

Assistance from the entire universe is needed if a person is to have both the psychic and the physical powers needed to live through the perils of earthly existence. While invocations like this one are ways of locating the human mode of being within the complex of powers throughout the universe, there are other ways in which the structure of human life and human society is formally coordinated with the movement of the cosmological order through the seasons.

Indeed the unity of the universe can be observed in the vast array of rituals carried out by diverse indigenous peoples around the planet as well as by the agricultural and literate societies, especially in their founding stages. The human cultural project in all its phases is authenticated by ritual insertion into the great seasonal sequence whereby Earth is constantly renewed. Thus a calendar of events is established and cosmic rituals are performed at their appropriate times.

Both personal and community affairs are validated by ritual integration of the human venture with the natural world. As with the cycle of the seasons, the periods of transformation in the cosmological order are celebrated by corresponding rituals in the human order. This is evident in the annual Thanksgiving celebration of the Haudenosaunee, or Iroquois, confederacy of the St. Lawrence River valley in North America. In their elaborate recitations these communities honor, through ritual acknowledgments that continue into the present, the beings layered throughout the natural world. For the Haudenosaunee the human community fulfills its roles in relation to all the various elements that sustain human life. While the winged creatures, four-legged creatures, insects, and all realms of life may be described as "spiritual" in our outsider effort to understand the ritual, what seems more important is that ongoing relationships are established and reaffirmed with these realms of being. Ritual relationships are acknowledged with the rain and the winds, with the mountains, the valleys, the plants, the animals, the homelands, the stars, and the sun in the heavens.

Personal intimacy with cosmic powers also finds expression among other peoples of North America such as the Siouan-speaking nations with their sweat-lodge ceremony and vision quest. The initiate, under the guidance of a shamanic personality, spends several days fasting and praying for the strength and orientation needed for future life. At this time the basic symbols that guide and protect a person are awakened in the consciousness of the initiate. A special name may be communicated that relates a person to some guardian life form. A song to be sung at the time of death may be received, a song that will enable the person to pass safely through the perils encountered at this time.

A remarkable example of intimacy with cosmic powers is found in the life story of Nicholas Black Elk, the Lakota *wichasha wakan*, or healer. When he was nine years old he had an astonishing vision relating to his specific historical situation and the destiny of his tribe. Special attention was given to the role of Black Elk himself in guiding the destiny of his people. In his vision he saw the six grandfathers representing the four directions, the sky, and Earth. At one moment in the prolonged vision, a great stallion in the heavens sang a song that rang out throughout the universe. In the words of Black Elk as narrated to John Neihardt:

> His voice was not loud, but it went all over the universe and filled it. There was nothing that did not hear, and it was more beautiful than anything can be. It was so beautiful that nothing anywhere could keep from dancing. The virgins danced, and all the circled horses. The leaves on the trees, the grasses on the hills and in the valleys, the waters in the creeks and in the rivers and the lakes, the four-legged and the two-legged and the wings of the wind—all danced together to the music of the stallion's song.[1]

If we look back to earlier, preindustrial times we see that the human relationship with the planet was different. There

---

[1] *Black Elk Speaks*, as told through John G. Neihardt by Nicholas Black Elk (Lincoln, NE: University of Nebraska Press, 2000 [orginally published in 1932]), 35; and Raymond J. De Mallie, ed., *The Sixth Grandfather: Black Elk's Teachings Given to John Niehardt* (Lincoln, NE: University of Nebraska Press, 1984), 133.

was never a time, certainly since the beginnings of the Neolithic agricultural period, when humans did not have some effect on the ecosystems of Earth. Yet there continued into the agricultural period a relationship whereby humans sought to authenticate their existence by entering into the cosmic cycle in the form of ritual celebrations. In this manner the ontological unity of the universe was recognized. Existence itself was celebrated. A pervasive consciousness of the sacred was established. Preservation of this relationship was a primary cultural motivation. Moreover, this was also an all-pervasive personal attitude that provided both cultural and individual identity with a personalized other. As suggested by Henri Frankfort in *Before Philosophy*, the natural world in the full diversity of its articulation was addressed as "thou," not as "it." The human community and the other life communities were all experienced as a single integral community of distinct personalities.

As we look back over the course of human affairs we find among the most significant and universal of human cultural developments is the effort to relate human affairs to the universe through ritual celebrations at certain moments in the annual sequence of the seasons, in the diurnal cycle, as well as in the life cycle of birth, maturity, and death. These events relating human affairs to the ever-recurring events of the natural world provide an authentication of the human mode of being. There is great significance in the coordination of human events, habitation, and communal architecture with the seasonal and astronomical cycles, because such coordination necessitates some awareness of time reckoning, of archaeoastronomy, and of a measured calendar. It led in some societies to precise observation of the phenomena of the sun and moon, the stars, the planets, and the natural phenomena.

In ancient China the cosmological order and human affairs were seen as existing in profound synchronicities and correlations. These relationships were so close that the entire range of political and social activity of the human world was arranged in accord with seasonal sequences in the natural world. Instructions in the meaning of the rituals are given in the liturgy book known as the *Book of Rites* (*Li Chi*). In this text the ritual observances for each season are prescribed. One entry gives the following warnings about resulting dilemmas if inappropriate seasonal rituals are performed:

> If in the first month of spring the governmental proceedings proper to summer were carried out, the rain would fall unseasonably, plants and trees would decay prematurely, and the states would be kept in continual fear. If the proceedings proper to autumn were carried out, there would be great pestilence among the people; boisterous winds would work their violence; rain would descend in torrents, mountain spinach, tufted grasses, ryegrass, and southernwood would grow up together. If the proceedings proper to winter were carried out, pools of water would produce destructive effects, snow and frost would prove very injurious, and the first sown seeds would not enter the ground.[2]

The object of all appropriately performed ceremonies is to bring down the spirits from above, and even the ancestors, in order to rectify relations between ruler, ministers, the bioregion, and the larger cosmos. The rituals maintain the generous feeling between father and son, and ensure the harmony between elder and younger brother. They adjust the relations between high and low, and give their proper places to husband and wife. The whole may be said to secure the blessing of Heaven. Humans are seen as the heart and mind of Heaven and Earth, and the visible embodiment of the five elements. They live in the enjoyment of all flavors, the discrimination of all notes, and the engagement with all colors.

In each one of these instances the integral relationship between the human and the universe was recognized and celebrated within a single community with plural expressions. The relationship was not primarily one of exploitation. The ritual

---

[2]*Book of Rites, Li Chi*, Book IV, Section I, part 11, "The Yueh Ling," trans. James Legge (New Hyde Park, NY: University Books, 1967), 257.

was a bonding of all the diverse components within a single, coherent whole and the purpose of the ritual was the flourishing of the entire range of beings within the universe. The understanding was that the human response was crucial for sustaining an integral functioning of all the diverse relationships throughout the bioregion, the cultural setting, and the larger cosmos.

In some societies, the annual ceremony was a reenactment of the creation of the universe. In Mesopotamia, for example, during a particular historical period the legend of Marduk and Tiamat was reenacted. Tiamat, the divine embodiment of the waters of chaos, was slain by the warrior deity, Marduk, and in their struggle the structure and diversity of the universe were established. In the Hindu tradition of India the patterns of ritual relationship of humans and the cosmos are expressed in the *Laws of Manu*. This book embodies ancient oral traditions that came down in several versions from the Vedic period when the Aryan peoples came into India around 1500 BCE. Eventually the traditions found literate expression in a single text. Here the acceptable pattern of human affairs was given in relation to cosmic order, or *rita*. Along with this, increasingly complex cosmological traditions were expressed in the early writings called the Vedas. The Vedas conclude in wisdom teachings called the *Upanishads* or Vedanta, the "end of the Vedas." Here the individuated self, or *atman*, was integrally linked to the larger universe in the form of the cosmic person, or *Brahman*. On every auspicious communal and individual occasion there was an effort to establish personhood at the center of the universe. The cosmic center, symbolized in script or icon or ritualized sound, such as *OM*, is where all things were validated, all things found their meaning and their security.

From this it becomes clear that in antiquity the primary cultural referent as regards reality, value, and power for religious expression is the universe in its full plurality and integrity. Every mode of being is universe-referent. Most particularly, with regard to the vast range of created beings that are culturally described and individually experienced, Earth is the immediate basis of cosmological reality and value. The question to explore, then, is in what ways did this change over time?

The Turn toward Historical Consciousness

Western civilization has experienced many and diverse ritual developments in the monumental civilizations of Mesopotamia and Egypt, Greece and Rome. Interacting with these larger civilizations, regional cultures, such as that of Israel, developed their own revelatory expressions. Thus, in the religion of Israel, the textual traditions associated with the Hebrew scriptures present elaborate ritual prescriptions evident in the Torah, or first five books of the Bible (Pentateuch), and especially in four of these books, Exodus, Leviticus, Numbers, and Deuteronomy. Many of the rituals are decidedly cosmological in their emphasis on seasonal celebrations related to animal husbandry and herding.

In the West, as Judaism and Christianity developed, the primacy of the universe shifted toward an increasing emphasis on historical events. This historical emphasis is clearly evident in the older Israelite religion, and especially in the Hebrew scriptures. A decisive moment in this regard occurred at Passover in the Exodus event that became the central historical and religious experience of Judaism. This complex and lengthy shift in the religion and thought of Israel was founded on the perceived historical intervention of divine deliverance from the servitude imposed on Israel during its enslavement period in Egypt. Deliverance of Israel at the Exodus was celebrated by the springtime cosmological festival, centered on Passover, but it was altered by a new emphasis on history. Passover becomes the first of a sequence of historical moments when divine communion with Israel is remembered as having occurred in place and time.

The immersion of the cosmos in the historical process was profoundly affected by a mythical-historical realism generated in the Hebrew Bible. For example, the Israelite priestly groups returning from exile in Babylon in the fourth century BCE measured creation itself in the historical passage of days. This sense of historical realism was reaffirmed with a creation narrative in which a numbered time-sequence of six days of creative generation was followed by a seventh day of ritual rest, or Sabbath. The Priestly text in chapter 1 of the Book of Genesis provided a clarity that had not been known previously in Israelite creation narratives such as that found in the Adam and Eve account of Genesis chapters 2 through 4. This time-precision with regard to creation set the framework for genealogical records and enabled the Israelite peoples to pull together in deeper historical and social identity.

The turn toward viewing divine intervention as occurring in history can be seen, for example, in the case of "judges," wise military leaders who were raised up at specific historical moments to assist Israel at times of need. So also the anonymous *nabi*, or bands of roving prophets, gave way to major named prophets who could see the divine hand in both cosmological happenings and historical events. One example is the prophetic visions of "the day of the Lord" at which time divine retribution would descend upon Israel for its transgressions. Described in mythic language and cast as a historical divination, these prophetic events occurred as particular, local happenings and were remembered as cosmic events in a sacred historical liturgy. Increasingly, the central liturgy of the Israelite people at the newly constructed temple in Jerusalem brought the cosmological vision of the psalms together with the prophetic focus on divinely ordained historical events.

This emerging historical consciousness was transmitted and incorporated into Christianity, especially in its emphasis on the incarnational event of Christ both as fulfilling the Jewish sense of sacred history and as giving rise to a new cosmological relationship. The historical realism woven into cosmological identity was emphasized by Saint John in his first epistle:

> Something which has existed since the beginning, that we have heard, and we have seen with our own eyes; and we have watched and touched with our hands, the Word, who is life—this is our subject. That life was made visible: we saw it and we are giving our testimony, telling you of the eternal life which was with the Father and has been made visible to us. What we have

seen and heard we are telling you so that you too may
be in union with us as we are in union with the Father
and with his Son Jesus Christ. (1 John 1:1–3)

A similar orientation, drawing attention to the physical
realism of the incarnational experience, is evident in a passage
of Saint Paul: "For just as the body is one and has many members, and all the members of the body, though many, are one
body, so it is with Christ. For by one spirit we were all baptized
into one body—Jews or Greeks, slaves or free—and all were
made to drink of one Spirit" (1 Corinthians: 12:12–13). And, in
a later passage, Paul asserts the significance of the physical risen
body of Christ saying, "Now if the dead are not raised then
Christ has not been raised, and if Christ has not been raised, you
are still in your sins. If our hope in Christ has been only for this
life, we are the most unfortunate of all people" (1 Corinthians
15:17–19). As regards the universe, then, Christ is considered to
satisfy the need for both individual and universe identity, since
Christ is considered the Cosmic Person through whom everything is sustained and given ultimate meaning.

The historical moment provides an underlying pattern for
the four Gospels, each of which presents the incarnation in a
distinctive way. The physical reality of Jesus, as we have noted
above, is emphasized by both John the Evangelist in the opening
verse of his first letter and by Saint Paul who in his Epistle to the
Corinthians asserts that the physical body of Jesus is truly risen.
The physical reality of the resurrection is emphasized. This
Christian focus on the physical reality of Christ is important; the
emphasis on the physical derives from the Jewish origins of
Christianity. The sense of the human soul, which was to become
a major metaphysical teaching of Christianity, was derived from
Greek Platonic philosophy whose influences were felt through-
out the Mediterranean region.

The emphasis on historical realism was given a special interpretation by Saint Augustine, bishop of Hippo in North Africa
in the fourth century CE, in his autobiography, *The Confessions*,
as well as in his masterwork, *The City of God*. Through these
works, and in the emerging Christian tradition, the story of the

universe began to lose its comprehensive mythic content in favor of subjective accounts of a devoted life and the realistic historical narratives of the Bible.

The emphasis on historical realism and the commitment to religious salvation through association with the narrative of the Gospels and the sequence of the redeemed community gave an even greater realism to the entire Christian process. Throughout this process the earlier cosmological emphasis on the primacy of the universe as the central reference for reality and value was being altered to a non-renewing human historical reference within an ever-renewing universe. Thus, as Christianity entered into its medieval period, it retained a strong cosmological strain in its liturgies and architectural expressions. But the historical emphases on Christ as having endured the passion and the cross gradually took precedence over the understanding of Christ as cosmic redeemer. These developments in religious consciousness were also tied to the growing understanding of Europe as a holy land.

Medieval Europe was the holy land of Christianity in its physical sense. That is, sacred sites of the older pagan peoples of Europe provided places where Christianity established itself as a spiritualizing force of missionary zeal. The thought expressed by the theologians of the period established a holy land of the mind fixed on mythic-historical re-creations of the Christ event. Thus, the passion and resurrection of Christ could be associated both with the actual sites in and around Jerusalem and with places of iconic representations of crucifixion, entombment, and deliverance from death. Liturgically, the Christian Eucharist was altered from its earlier sense of a communal *agape*, or meal of the beloved community, into a remembrance of the historical passion of Christ who atoned for the sin of humanity. Simultaneously, the cosmological patterns evident in religious buildings and sacred symbols endured, but grew weaker in significance as the subjective experience of divine intervention in history grew stronger.

One important thinker in this medieval period was Thomas Aquinas who proposed that the universe be taken as the primary referent in any serious discussion of the order of the universe. In his *Summa Contra Gentiles* (Book II, chapter 26) he says, "The order of the universe is the ultimate and noblest perfection in things." In his *Summa Theologica* (I, q. 47, a. l.), in answer to the question of why is there is such a vast diversity of things, Aquinas says that "Because the Divine could not manifest itself fully by creating another deity, he created the great diversity of things so that the perfection lacking to one would be supplied by the others and that the whole universe together would participate in and manifest the divine more than any single being whatsoever."

Both these quotations are supported by the passage from the first story of creation found in the Book of Genesis. After each day of creation God says that it is good, but after the last day, looking at the integral relation of things, he says that it is very good. In Christianity this would seem to indicate that whatever is done to heal some primordial fault in the universe by way of incarnation and redemption, that too would ultimately and most properly be intended to heal the entire universe in its wonder and its splendor.

As we have observed, the historical challenge to the cosmological worldview that we find in the Gospels was initiated at the time of the Exodus, when a cosmological experience of the divine was transformed into a historical experience. This might be considered the foundation of the importance of the sense of historical realism in Western civilization. Moreover, the commitment to the superiority of the human over the natural world set in motion those forces of alienation that would eventually lead in the West to our strong sense of anthropocentrism.

Still, the celebration of creation in medieval Christianity retained its place in the visions, the paintings, the writings, the songs, and the liturgies of many persons and religious communities. The natural world was exalted as containing modes of divine presence. We find expressions of this in forms of Celtic animism, in Benedictine stewardship liturgies, in the devotions of Hildegard of Bingen, in the fraternal prayer of Francis of Assisi, and, in the contemporary period, in the evolutionary model of Teilhard de Chardin. The greater presence of the universe, celebrated in dawn and evening observances and in the seasonal cycles, was the central liturgy as ordered in accord with the movement of the heavens. Later, the French Victorines and the German Rhineland mystics expressed a Neo-Platonic devotion to vestiges of God in the natural world. In their diverse ways they sought an emptying of consciousness, especially of all images, while still exalting the world of natural forms.

It was against this background that the realism of the modern scientific inquiry emerged. With the Renaissance shattering of the older Ptolemaic map and design of the universe, the historical emergence of the universe in its structure and functioning and the place of the human in this context began to be dominant. Here is where the alienation of the human from the natural world became a new religious vision and a new economic possibility. The bonding with the universe that had survived for fifteen centuries in the Christian world began to be undermined. From modest beginnings the scientific understanding of the universe would radically alter religious perspectives. Sciences such as geology and physics, which had very little to do with each other, began to develop and to establish new ways of thinking about the known world.

Throughout Europe and in the newly discovered worlds of the Americas, Asia, and the Pacific, religious rituals continued to meld these two great drives in human consciousness, the historical and the cosmological. It was not difficult for European Christians to move out over the planet in a geographic extension of the mythic connection between Europe and the Holy Land centered on Jerusalem. However, there was another difficulty as European and Euro-American peoples began to shape a new world of the mind completely alien from the consciousness formed by the Greek and Latin heritage of the Mediterranean world. In succeeding centuries of encounter with diverse bioregions and peoples, the Western mind became thoroughly absorbed in the discovery it was making of New Worlds.

After Renaissance thought had turned toward classical patterns of order and rational design, the developments now described as the Enlightenment eventually gave rise to a science that perceived an emergent universe and the evolving life forms of Earth. This evolutionary story of life on Earth came to shape people's sense of the universe throughout the literate world. In effect, science rather than a sense of salvation history came to dominate cosmological thinking. The religions, for the most part, maintained ancient cosmological symbols in their liturgies, but effectively retreated from cosmological knowledge of reality. These two worlds, namely, the world of historical change and the world of the sacred cosmos, became two separate realms with little rapport with each other.

In many religious settings, sacred teachings became preaching about the sacraments and the sacred calendar. These teachings remained as they had been throughout the earlier Christian centuries. Belief statements and catechetical training also remained cast in an earlier terminology. A residual scholastic thought that endured from the medieval period did take note of the cosmological relationship of the world to the creative energies of the divine. However, after Thomas Aquinas, Christianity did not produce a thinker until Teilhard de Chardin who could integrate Christian emphasis on salvation history with Christian attention to the incarnational sense of the sacred in matter.

Thus, the Western religious traditions all continued their centuries-old manners of salvific quest expressed in liturgies that often had powerful cosmological symbols. One such biblical idea, that of a people chosen by God in relation to land, lost its cosmological anchor and became an affirmation of humans apart from Earth. The particularization in the biblical stories of the chosen people was transmuted into new religious and political ideas. For example, the term "world" was often considered to carry derogatory connotations. Humans tended to think of themselves as blessed precisely because they were distinct from other creatures and had the intellectual capacity for self-reflection. Moreover, having been given the revelation of the Bible and having been called to share the blessings that come through the Christian faith, many religious communities began to think of themselves as composing an elite community of the saved, forever separated from those not saved.

In this context we need to reexamine the entire question of "chosenness." Christianity has lived through the centuries with a sense of being particularly blessed with a saving belief in a redemption brought about through the special presence of the divine in human form. In this sense, Christianity's experience of being chosen has been a major force in establishing Western civilization. Nevertheless, it seems that a Second Exodus has been in the making since the time of Copernicus. Is it possible that we are being called to make an Exodus out of one sacred community locked in its historical self-salvific identity into the larger sacred community of life? Can we experience an Exodus from a mode of oppression that we humans have become to the planet to a form of enhancement for the Earth community?

A Second Exodus

One of the basic insights of the historian Eric Voegelin, in his work entitled *Israel and Revelation*, is that of a Second Exodus. Israel, after the primary Exodus from Egypt to the Promised Land, needed to return to Egypt in a second Exodus to bring the salvation experience in Jerusalem back into the world from which Israel had been delivered. Salvation was not to be contained simply in any limited portion of the human community. This responsibility was a challenge experienced by the Christian community with its sense of universal evangelization.

The dynamic of Christianity that has insisted on redemption out of the original world of nature into the sacred world of grace should now lead us to return to the natural world to bring a new understanding for the entire planet. This could be an explanation for the remarkable scientific venture of the past few centuries. This return has not yet brought Christianity into full relationship with the natural world because we have not fully embraced modern science and its understanding of evolution.

The earlier modes of understanding the universe and celebrating planet Earth involved thinking in terms of an ever-renewing seasonal sequence. But then more precise observation by Copernicus, Galileo, and Newton provided new ways of thinking about the universe. With the thought of René Descartes and Francis Bacon, a broadened intellectual worldview emerged, one that gradually reimagined itself in terms of a much older process. Humans moved from the ever-renewing world of Ptolemy to the cosmological world of Copernicus. We might think of this process as a Second Exodus made possible by science.

The Christian world of our times needs to make its own Second Exodus out of its medieval Ptolemaic background into the modern world by adopting the story of the universe as this is now available to us. This does not involve abandoning the ever-renewing world of historical times. The planet will always pass through its ever-renewing cycles of the seasons. We will remain grounded in the Christian context of renewal as this existed before the empirical sciences began their discoveries. The evolutionary and the cyclical do not negate each other. They only need to integrate each other.

To do this will require the adoption of liturgies based on the universe story from its beginning some fourteen billion years ago throughout its sequence of transforming episodes whereby the universe and Earth have come into existence. One of the great differences in our present sequence of liturgies is that traditional liturgies have been adopted into the ever-renewing seasonal cycles. These will necessarily continue even while we move into celebration of the evolutionary transformation moments, which might find annual expression of one-time events just as the life story of Christ narrates one-time events that are now celebrated annually.

This Second Exodus will bring about a deep psychic transformation in the Christian world. The principle of comprehensiveness is central to the sacred story. Absorption of the universe into the sacred realm of the human and absorption of the human into the sacred realm of the universe is required for any satisfactory way into the future. Only when we begin to think of the emergent universe as the comprehensive realm of the sacred will we be able to overcome our present assault on the universe in its Earth manifestation. Signs of this Second Exodus out of our destructive industrial mode are already observable.

An Emerging Sense of Limits

After World War II there was evidence that the industrial world of the human was wreaking havoc throughout the natural world. Observing this situation, Aurelio Peccei, an official in the Fiat Automobile Corporation, toured the world to consult with the most competent persons he knew of to seek out some effective way of dealing with the situation. The result was a call to Rome of some forty persons to discuss ways of dealing with what came to be called the "Global Problematique," that is, the issue of shaping a viable mode of human presence on the planet.

After a meeting in 1968 this group decided to call itself the Club of Rome. The group commissioned a number of studies of industrial production processes and their consequences to local bioregions and to Earth as a whole. They explored what might be done to shape a more viable future for human relationships with the planet. The first of these studies, published in 1972 under the title *Limits to Growth*, shocked the world at that time. The dissemination of the ideas contained in the study marked a decisive moment in human awareness of what was happening to Earth and how a disastrous future could be avoided. That same year, the Stockholm Conference on environmental issues was convened by the United Nations. It was the first time that the various nations of the world gave official recognition to the disturbed situation of the planet Earth. One of the results of the conference was that most of the nations represented there subsequently established the first national environmental protection agencies.

These developments give some sense of the initial attempts to awaken a sense of urgency and to establish a more viable rapport between the human community and the more comprehensive Earth community. However, in the decades since these events took place, while some progress has been made, there has not been sufficient action taken by the four basic human establishments: the political, economic, educational, and religious communities. They appear to be either incompetent or unwilling to consider the magnitude of the ecological problems we are facing. All four establishments are failing for the same reason. They all place a discontinuity between the human and the other-than-human modes of being. They give the basic values and rights to the human. No basic values and rights are given to non-human modes of being. The human is the primary referent in all questions, whether of law, economics, education, or religion. This question of the primary referent must be the basic issue up for discussion. Until this question of the human as our basic referent is reconsidered, there is no adequate way to proceed.

At present there is a devastating relationship between the human community and the Earth community. That humans, with all their intelligence, should be so destructive is something of an anomaly. Yet the disruption that is being effected throughout the planet is occurring through human agency. A remedy needs to be found.

The destruction is the result of scientific technologies whereby the resources of Earth are being exploited for human benefit without regard for the consequences on the life systems of the planet. A mystique of use as our primary relationship with the planet has been developed and is being deployed with ruthlessness beyond understanding.

This represents a great contrast with the earliest phase of human development when Earth and the solar system, with its planets and the stars beyond, were considered to be manifestations of sacred, even divine powers. In this period the human project was validated by ritual celebration of those great moments of transition in the ever-renewing seasonal cycle when the numinous world was experienced as especially present to the human community.

The entire universe was experienced as a cosmic liturgy. There was the daily sequence when at dawn and at sunset the numinous presence was more available to human perception. Then there was the yearly cycle. The annual cycle of death and rebirth occurred as the sun declined to its nadir at the winter solstice and then was reborn in the springtime when a renewal of life came forth with dramatic manifestations. This was an ever-renewing seasonal sequence. What was born, died. What died was born again. If there was a never-ending tendency toward death, there was also the inevitable life-renewal.

The liturgy of the Christian world was woven over the years into cosmological patterns. Liturgical celebrations were coordinated with the hours of the day, especially with the dawn and sunset, as well as with the seasons of the year. The nativity was coordinated with the solstice. The resurrection was celebrated at the time of the spring equinox.

Much of this could not easily survive the series of later scientific discoveries of the structure and movements of the formerly astrological bodies. The change of consciousness and mental commitment needed to move from a direct divine disposition of the universe at the moment of creation with Earth at the center was too demanding. Moreover, it was a shock to the ancient cosmological understanding that undergirded the historical worldviews upon which political power rested. It was especially challenging to traditional religious thought when scientific discoveries began indicating that the planet Earth was much older than the five thousand years indicated by calculations based on a reading of the Bible. Scientific insights led to an alienation from the former sense of the cosmos as the locus for the meeting of the divine and the human. This became especially clear as science went on to discover that the universe was a self-emergent and a self-organizing process.

At this time the sacred character of the universe was overshadowed and even lost. The universe was seen as the expression of random forces which had evolved over a vast period of time. While the sense of the universe as sacred had been diminished with the Christian focus on the presence of the divine reality in the person of a human being, a cosmological underpinning provided stability, especially in the deep psychic commitments of Christian liturgy.

During the modern period, however, the Christian faith has lost a significant part of its vigor. Its mission to establish a divine presence within the human community has become exceedingly difficult. The current scientific understanding of the universe and of human life does not seem to welcome a religious interpretation. The devastation of Earth we are experiencing and the loss of a sense of the sacred seem to be profoundly related. Just how this sense of the sacred can be restored requires a new depth of understanding on the part of the Christian world. Unless a sense of the sacred is restored, the doctrine of use embedded in our economic system will continue to devastate the planet.

My suggestion is that just as Christianity in its developing phase established itself in intimate relations with the structure and functioning of the universe in its liturgical processes, so now there is a need to adopt a new sense of a self-emergent universe as a sacred mode whereby the divine becomes present to the human community. If an ever-renewing universe was what first emerged in human consciousness, this need not be negated, since this is one of the most basic patterns of the solar system and of the planet Earth. The solstices and the equinoxes will remain dominant experiences for all living forms on Earth.

With regard to the new cosmic-based rituals that will be involved, we might consider the moment of emergence of the first photons, the primary movements of differentiation, inner-spontaneity, and the bonding of all things with each other. Then we might consider the formation of the galaxies, as well as the emergence of the first molecules, microbes, cells, and animals.

We need to think of the universe in its comprehensive dimensions, for indeed the universe is the only self-referent mode of being in the phenomenal world. Every other mode of being is universe-referent in its structure and in its functioning.

We need to understand that the purpose of the universe is not focused on any one single being, but depends upon the entire multiplicity of beings. This view accords with that indicated in the first creation story in the Book of Genesis, where after each day of creation God says that it is good, but after the last day of creation he sees the world complete and so he says that it is very good.

It follows from these reflections that whatever is done by the power that brought the universe into being is done primarily for the perfection of the entire creation, not for any individual within creation. Thus revelation, incarnation and redemption are primarily for the entire universe, not primarily for any group or individual being within the universe.

Acceptance of the universe as a self-emergent evolutionary process is one of the greatest challenges faced by Western civilization in establishing an integral intellectual and religious life orientation. Yet it is critical, in the light of present data, to embrace the science of evolution in our efforts to establish an acceptable cosmology. Although we have had such amazing scientific insight into the structure and functioning of the planet, we have been without any meaningful interpretation of the data. We have therefore lost whatever intimacy we had previously with the surrounding universe.

What is proposed here as a necessary development is to once again establish an awareness of the unity of human affairs with the functioning of the universe and Earth. This unity finds expression in the threefold cosmological cycles: daily, seasonal, and planetary. There is the daily cycle that passes from sunrise to sunset, the dawn and dusk bringing the community out of the mystical period of awakening consciousness when the dreams of the preceding night are narrated for the guidance that they might offer to the community. Then there is birth, maturity, and death in the seasonal cycle of each living form, reflected in the winter solstice, the springtime renewal, the summertime fullness, the autumnal harvest. Finally there is the planetary cycle of Earth's orbit around the sun. While no society had its full expression of these various moments of grace in which the human communed with the cosmos, they each had their intimacy with the larger community of life. Now we are called to our Second Exodus, our return to the universe as vibrant cosmic liturgy.

Appendix

Reinventing the Human at the Species Level

The present human situation can be described in three sentences:

In the twentieth century the glory of the human has become the desolation of Earth.

The desolation of Earth is becoming the destiny of the human.

All human institutions, professions, programs and activities must now be judged primarily by the extent to which they inhibit, ignore, or foster a mutually enhancing human-Earth relationship.

In the light of these statements it is proposed that the historical mission of our times is:

> To reinvent the human
> At the species level
> With critical reflection
> Within the community of life systems
> In a time-developmental context
> By means of story and
> Shared dream experience.

The first phrase, "to reinvent the human," suggests that the planetary crisis we are facing seems to be beyond the competence of our present cultural traditions. What is needed is something beyond existing traditions to bring us back to the most fundamental aspect of the human: giving shape to ourselves. The issue has never been as critical as it is now. The human is at an impasse because we have brought the entire set of life systems of the planet to an impasse. The viability of the human is in question.

Our present difficulty is that we envisage the universe simply in its physical dimensions. We have lost the awareness that the universe has from the beginning been a psychic-spiritual as well as a material-physical reality. It has taken the entire course of the evolutionary process for the universe to find its expression in the florescence of living forms and in the various modes of consciousness that are manifested throughout Earth.

The immense curvature of space holds all things together in an embrace that is sufficiently closed to provide structural integrity to the universe and yet sufficiently open to enable the universe to continue its unfolding. Within this context we need a new appreciation of our cosmocentric identity.

Second, we must work "at the species level" because our problems are primarily problems of species. This is clear in every aspect of the human. As regards economics, we need not simply a national or a global economy, but a species economy. Our schools of business teach the skills whereby the greatest possible amount of natural resources is processed as quickly as possible, put through the consumer economy, and then passed on to the junk heap where it is at best useless and at worst toxic to every living being. There is need for the human species to develop reciprocal economic relationships with other life forms, providing a sustaining pattern of mutual support, as is the case with other life systems.

As regards law, we need a species legal tradition that would provide for the legal rights of geological and biological as well as human components of Earth community. A legal system exclusively for humans is not realistic. Habitat, for example, must be given legal status as sacred and inviolable.

Third, I say "with critical reflection" because this reinventing of the human needs to be done with utmost competence. We need all our scientific knowledge. We cannot abandon our technologies. We must, however, ensure that our technologies are coherent with the technologies of the natural world. Our knowledge needs to be a creative response to the natural world rather than a domination of the natural world.

We insist on critical understanding as we enter the Ecological age in order to avoid a romantic attraction to the natural world that would not meet the urgencies of what we are about. The natural world is violent and dangerous as well as serene and benign. Our intimacies with the natural world must not conceal the fact that we are engaged in a constant struggle with natural forces. Life has a bitter and burdensome aspect at all levels, yet its total effect is to strengthen the inner substance of the living world and to provide the never-ending excitement of a grand adventure.

Fourth, we need to reinvent the human "within the community of life systems." Because Earth is not adequately understood either by our spiritual or by our scientific traditions, the human has become an addendum or an intrusion. We have found this situation to our liking since it enables us to avoid the problem of integral presence to Earth. This attitude prevents us from considering Earth as a single society with ethical relations determined primarily by the well-being of the total Earth community.

But while Earth is a single integral community, it is not a global sameness. It is highly differentiated in bioregional communities—in arctic as well as tropical regions, in mountains, valleys, plains, and coastal regions. These bioregions can be described as identifiable geographical areas of interacting life systems that are relatively self-sustaining in the ever-renewing processes of nature. As the functional units of the planet these bioregions can be described as self-propagating, self-nourishing, self-educating, self-governing, self-healing, and self-fulfilling communities.

Human population levels, our economic activities, our educational processes, our governance, our healing, our fulfillment must be envisaged as integral with this community process. Earth itself is the primary progenitor, economist, educator, lawgiver, healer, and fulfillment for everything on Earth.

There are great difficulties in identifying just how to establish a viable context for a flourishing and sustainable human mode of being. Of one thing we can be sure, however, and it is that our own future is inseparable from the future of the larger life community. That is because this life community brought us into being and sustains us in every expression of our human quality of life—in our aesthetic and emotional sensitivities, our intellectual perceptions, our sense of the divine, and our physical nourishment and our bodily healing.

Fifth, reinventing the human must take place in "a time-developmental context." This constitutes what might be called the cosmological dimension of the program we are outlining here. Our sense of who we are and what our role is must begin where the universe begins. Not only the formation of the universe but also our own physical and spiritual shaping begin with the origin of the universe.

The ethical formation required is governed by three basic principles: differentiation, subjectivity, and communion.

Our present course is a violation of each of these three principles in their most primordial expression. Whereas the basic direction of the evolutionary process is toward constant differentiation within a functional order of things, our modern world is directed toward monocultures. This is the inherent direction of the entire industrial age. Industry requires a standardization, an invariant process of multiplication with no enrichment of meaning. In an acceptable cultural context, we would recognize that the unique properties of each reality determine its absolute value both for the individual and for the community. These are fulfilled in each other. Violation of the individual is an assault on the community.

As a second ethical imperative derived from the cosmological process, we find that each individual is not only different from every other being in the universe but also has its own inner artic-ulation. Each being in its subjective depths carries the numinous mystery whence the universe emerges into being. This we might identify as the sacred depth of the individual, one's subjectivity.

The third ethical imperative of communion reminds us that the entire universe is bonded together in such a way that the presence of each individual is felt throughout the entire spatial and temporal range of the universe. This capacity for bonding of the components of the universe with each other enables the vast variety of beings to come into existence in that gorgeous profusion that we observe about us.

From this we can appreciate the directing and energizing role played by "the story of the universe." This story that we know through empirical observation of the world is our most valuable resource in establishing a viable mode of being for the human species as well as for all those stupendous life systems whereby Earth achieves its grandeur, its fertility, and its capaci-ty for endless self-renewal.

This story, as told in its galactic expansion, its Earth forma-tion, its life emergence, and its manifestation of consciousness in the human, fulfills in our times the role of the mythic accounts of the universe that existed in earlier times when human aware-ness was dominated by a spatial mode of consciousness. The story represents a transition in human awareness from the uni-verse as cosmos to the universe as cosmogenesis. It represents a shift in the spiritual path from a mandala-like journey to the center of an abiding world to the great irreversible journey of the universe itself as the primary sacred journey. This journey of the universe is the journey of each individual being in the uni-verse. So this story of the great journey is an exciting revelatory story that gives us our macrophase identity—the larger dimen-sion of meaning that we need. To be able to identify the microphase of our being with the macrophase mode of the uni-verse is the quintessence of what needs to be achieved.

The present imperative of the human is that this journey con-tinue on into the future in the integrity of the unfolding life sys-tems of Earth, which presently are threatened in their survival.

Our great failure is the termination of the journey for so many of the most brilliant species of the life community. The horrendous fact is that we are, as the scientist Norman Myers has indicated, in an extinction spasm that is likely to produce "the greatest single setback to life's abundance and diversity since the first flickerings of life almost four billion years ago."[1] The labor and care expended over some billions of years and untold billions of experiments to bring forth such a gorgeous Earth is being negated within less than a century for what is considered "progress" toward a better life in a better world.

The final aspect of our statement concerning the ethical imperative of our times is "the shared dream experience." The creative process, whether in the human or the cosmological order, is too mysterious for easy explanation. Yet we all have the experience of creative activity. Since human processes involve much trial and error with only occasional success at any high level of distinction, we may well believe that the cosmological process has also passed through a vast period of experimentation in order to achieve the ordered processes of our present universe.

In both instances something is perceived in a dim and uncertain manner, something radiant with meaning that draws us on to a further clarification of our understanding and our activity. Suddenly out of the formless condition a formed reality appears. This process can be described in many ways, as a groping, or as a feeling, or as an imaginative process. The most appropriate way of describing the process seems to be that of dream realization. The universe appears to be the fulfillment of something so highly imaginative and so overwhelming that it must have been dreamed into existence.

But if the dream is creative we must also recognize that few things are so destructive as a dream or entrancement that has lost the integrity of its meaning and entered into exaggerated and destructive manifestation. This has happened often enough

---

[1]Norman Myers, "The Biodiversity Crisis and the End of Evolution," *The Environmentalist* (1996): 37–47.

with political ideologies and with religious visionaries, but there is no dream or entrancement in the history of Earth that has wrought the destruction that is taking place in the entrancement with industrial civilization. Such entrancement must be considered as a profound cultural pathology. It can be dealt with only by a correspondingly deep cultural therapy.

Such is our present situation. We are involved not simply with an ethical issue but with a disturbance sanctioned by the very structures of the culture itself in its present phase. The governing dream of the twentieth century appears as a kind of ultimate manifestation of that deep inner rage of Western society against its earthly condition. As with the goose that laid the golden egg, so Earth is assaulted in a vain effort to possess not simply the magnificent fruits of Earth, but the power itself whereby these splendors have emerged.

At such a moment a new revelatory experience is needed, an experience wherein human consciousness awakens to the grandeur and sacred quality of Earth process. This awakening is our human participation in the dream of Earth, the dream that is carried in its integrity not in any of Earth's cultural expressions but in the depths of our genetic coding. Therein Earth functions at a depth beyond our capacity for active thought. We can only be sensitized to what is being revealed to us. Such participation in the dream of Earth we probably have not had since earlier times, but therein lies our hope for the future for ourselves and for the entire Earth community.

Index

Also Published in the Ecology and Justice Series